Food Science, Safety and Quality Control

Food Science, Safety and Quality Control

Edited by **Margo Field**

New York

SYRAWOOD
PUBLISHING HOUSE

Published by Syrawood Publishing House,
750 Third Avenue, 9th Floor,
New York, NY 10017, USA
www.syrawoodpublishinghouse.com

Food Science, Safety and Quality Control
Edited by Margo Field

International Standard Book Number: 978-1-68286-013-7 (Hardback)

Printed in the United States of America.

Contents

Preface

Every book is initially just a concept; it takes months of research and hard work to give it the final shape in which the readers receive it. In its early stages, this book also went through rigorous reviewing. The notable contributions made by experts from across the globe were first molded into patterned chapters and then arranged in a sensibly sequential manner to bring out the best results.

Food safety and quality control are two very important aspects of the food industry. This innovative and comprehensive book integrates the well-developed theory and practical applications of food science with the concepts of quality control and safety standards that are practiced in the industry. Food toxicants, contaminants, mycotoxins, nutrition, food processing and control technologies, etc. are some of the topics that have been discussed in detail. It is a complete source of knowledge on the present status of food science. With state-of-the-art inputs by acclaimed experts of this field, this text targets students and professionals.

It has been my immense pleasure to be a part of this project and to contribute my years of learning in such a meaningful form. I would like to take this opportunity to thank all the people who have been associated with the completion of this book at any step.

Editor

The Effect of Extrusion Conditions on the Physicochemical Properties and Sensory Characteristics of Millet – Cowpea Based Fura

K. B. Filli[1*], I. Nkama[2], V. A. Jideani[3] and U. M. Abubakar[2]

[1]Department of Food Science and Technology Federal University of Technology Yola, PMB 2076 Yola, Adamawa State, Nigeria and SIK – The Swedish Institute for Food and Biotechnology Gothenburg, Sweden.
[2]Department of Food Science and Technology University of Maiduguri PMB 1069 Maiduguri, Borno State, Nigeria.
[3]Department of Food Technology, Cape Peninsula University of Technology, Bellville 7535, Cape Town, South Africa.

ABSTRACT

A three-factor three level Response surface methodology central composite retortable design (CCRD) was adopted to study the effect of feed composition (X_1), feed moisture content (X_2) and screw speed (X_3) on proximate compositin, amino acid and sensory evaluation during extrusion of pearl millet and cowpea flour mixtures for the purpose of fura production. The mean observed value of protein for the fura extrudates ranged from 11.2 – 16.8%. Analysis of variance indicates that linear and quadratic effects significantly ($P<0.05$) affected the protein content of fura extrudates as expected. The mean value of lysine for the extrudates ranged from 5.1 - 6.6g/100g protein and the methionine content ranged from 1.3 - 3.8g /100g protein. The regression models fitted to the experimental data showed high coefficients of determinants with R^2 = 0.96, 0.94, 0.94, 0.85 and 0.80 for protein (CHON), carbohydrate (CHO), fat (FAT), ash (ASH) and water (HOH) respectively. The R^2 = were 0.90, 0.85, 0.86, 0.92, 0.88, 0.85 and 0.93 for lysine, i/leucine, leucine, valine, methionine – cystine, threonine and tryptophan respectively. The coefficients shows good fit. The importance of process variables on system parameters and physical properties could be ranked in the following order: Feed Composition (X_1)>Feed Moisture (X_2)>Screw Speed (X_3). The optimum values obtained for feed composition, feed moisture and screw speed are 36.5% cowpea level, 22.3% feed moisture content and 186.7 rpm

*Corresponding author: Email: kbfilli@yahoo.com; kbfilli@gmail.com

respectively. The corresponding optimum values of lysine, protein content, expansion ratio and bulk density are 6.6g/100g protein, 16.6%, 2.8 and 0.52 Kgm^{-3} respectively. The data obtained from the study could be used for control of product characteristics and possible projection for the commercial production of fura or any enriched protein based food from the blends pearl millet and cowpea.

Keywords: Millet; cowpea; extrusion; fura; protein; amino acid; feed composition.

1. INTRODUCTION

Fura is a semi-solid dumpling cereal based meal (Jideani et al., 2002). It is produced principally from millet or sorghum flour blended with spices, compressed into flour balls and boiled for 30 minutes (Jideani et al., 2002). The mode of processing may vary slightly among different communities, but the basic ingredient remains the same. The product lacks process specifications governing composition, ingredients, additives and shelf life. Processing has remained a home-based or artisanal activity that is carried out with rudimentary equipment and techniques. Depending on the community fura is usually consumed after mashing with *nono* (local yoghurt) or mashed in water before consumption in the form of porridge (Filli et al., 2010).

Extrusion technology is one of the contemporary food processing technologies applied to foods (Harper and Jansen, 1985) and can be applied to mitigate the problems associated with processing of traditional cereal based products in terms of improvement in functionality, physical state and shelf stability. It offers many advantages over other process technologies in terms of, preparation of ready – to – eat foods of desired shape, size, texture and sensory characteristics at relatively low processing cost (Sumathi et al., 2007). Extrusion is a powerful food processing operation, which utilizes high temperature and high shear force to produce a product with unique physical and chemical characteristics (Pansawat et al., 2008). Excess water is not available in extrusion and the starch granules do not swell and rapture, as in classical gelatinization, but are instead mechanically disrupted by high shear forces and drastic pressure changes resulting in disappearance of native starch crystallinity, plasticization, expansion of the food structure, reduced paste viscosity, loss of water holding, increased reconstitutability of the extrudate, softer product texture and changes in colour (Onyango et al., 2004).

Pearl millet (*Pennisetum glaucum* [L.] R.Br.), is grown extensively in the dry areas of western and southern India and along the West African sub region where it is used as food for an estimated 400 million people (Hoseney et al., 1992). Pearl millet is an important cereal, contributing to the calorie and protein requirements of people in the semi-arid tropics (SAT). It is grown mostly in regions of low rainfall and is capable of withstanding adverse agro climatic conditions. More than 80% of the production is used for human consumption, particularly in the SAT region of Africa and Asia. Several food preparations are made from pearl millet in Africa and India (Vogel and Graham, 1979). Many countries in the developing world have become heavily dependent on imported staple foods which conditions for local production are poor or non-existent making the demand for traditional based products not attractive. In Nigeria pearl millet has remained as a staple food in the form of gruels for the poor especially in the northern part of the country. Limited efforts has been made by the scientific community to diversify its food uses by the application of modern technology to upgrade the traditional methods of contemporary food processing technology for millet utilization, despite the advancement in scientific research. The application of a contemporary technology for the traditional products in our fast growing social environment can enhance the development and acceptability of indigenous traditional based foods. Filli et al. (2010) reported that the extrusion cooking of 'fura' from blends of millet – soybean constituted a great improvement on the shelf stability of traditional fura that is usually at high moisture content of between 60 - 75% which readily deteriorates on storage at room temperature when compared with fura extrudates obtained that had moisture content less than 7 g/100g which did not require refrigeration for storage.

The cowpea is indigenous to Africa which is consumed in various forms in different African countries. The cowpea is probably the most popular grain legume in West Africa. Cowpeas like other starchy legumes are an important source of protein and B vitamins in developing countries. The common bean is an important food on a world basis and provides significant amounts of proteins, calories, minerals and vitamins to many human populations. Beans are important constituents of the diet and provide economical sources of proteins and energy. They have also been indicated to show apparent benefits of soluble fiber in preventing heart disease. However, the long cooking time required and presence of anti-nutritional substances in the whole bean limit their use. In view of the increasing population growth in developing countries, the production and utilization of grain legumes will have to be increased. Improved productivity, availability and utilization of grain legumes in the diet of people can help to alleviate the protein-calorie malnutrition problem prevalent in most developing sub Sahara African countries. Although the chemical composition of cowpea has been reported in some publications (Mwasaru et al., 1999), little information is available on the extrusion of millet – cowpea flour mixtures. A review of the available literature reveals that more effort has been invested in the nutritional and chemical evaluation of these legumes rather than applications of processing technology to it. According to Gujska et al. (1996), extrusion cooking has good potential for making desirable forms of beans economically available in developing countries. Nontraditional methods of processing cowpea such as thermal extrusion are needed for expanded utilization of dry edible beans. It is well known fact that addition of legumes to cereals increases both content and quality of protein mix (Obatolu, 2002). Wu et al. (2010) reported the inclusion of flaxseed to maize to improve the protein content and quality. Therefore, with respect to lysine and the sulphur containing amino acids, legume and cereal proteins are nutritionally complimentary. The inclusion of cowpea as a basic ingredient in producing fura through extrusion can improve its physical state and functionality. The poor quality of protein and high viscosity of traditional fura gruel makes it difficult to consume enough quantity to meet both energy and protein requirements. Nkama and Filli (2006) and Filli and Nkama (2007), reported that extruded fura from cereal legume blends provided consumers with a fast, easy way to prepare nutritious fura which is similar to the traditional fura. Extrusion enhanced the water uptake of the product, with reduction in viscosity which is an indication of concomitant increase in nutrient density, but the process method was not optimized.

Response surface method (RSM) is a statistical – mathematical tool which uses quantitative data in an experimental design to determine, and simultaneously solve multivariate equations, to optimize processes or products (Sefa – Dedeh et al., 2003); it has been successfully used for developing, improving and optimizing processes (Wang et al., 2007).

The objectives of this work was to study the effect of process conditions (feed composition, feed moisture and screw speed) on the physicochemical properties of fura from pearl millet and cowpea flour mixtures using response surface methodology.

2. MATERIALS AND METHODS

2.1 Flour Preparation from Pearl Millet

The process of flour preparation consists of dry cleaning of millet i.e. winnowing using an aspirator Vegvari Ferenc (OB125, Hungary). The kernels were thereafter dehulled after mild wetting of the grain using a rice dehuller (India) at the Jimeta Main Market, Yola, Nigeria. After dehulling, the grains were washed and then dried in a Chirana convection oven model (HS 201A, Czech Republic) at 50°C for 24 h to 14% moisture content. The dried grain was milled using a Brabender roller mill (OHG DUISBURG model 279002, Germany) equipped with a 150 μm screen.

2.2 Flour Preparation from Cowpea

The cowpea was steeped in tap water at 28°C for a period of 30 minutes to loosen the seed coat in a plastic bowl. This was followed by decorticating using pestle and mortar made of wood. The kernels of decorticated mass was dried at 50°C to approximately14% moisture content in a Chirana convection oven model (HS 201A, Czech Republic) for 24 hours. The grain mass was winnowed to remove the hulls and other lighter materials using an aspirator Vegvari Ferenc (OB125, Hungary). The winnowed cowpea kernels were ground in a laboratory disc mill (made in Nigeria) to fine flour. The flour obtained was sieved using a 150μm screen size Brabender (OHG Duisburg type, Germany) and the underflow was used for extrusion after blending with millet.

2.3 Spice Preparations

Kimba (Negro pepper) and ginger were sorted and cleaned manually before drying in a Chirana convection oven model (HS 201A, Czech Republic) at 60°C for five h. The seeds were then milled by pounding using pestle and mortar. The ground mass was sieved using a 150 μm screen size.

2.4 Blend Preparations and Moisture Adjustment

Millet flour (M_F) and cowpea flour (C_F) were mixed at various weight ratios, and the total moisture contents of the blends adjusted to the desired values with a mixer as described by Zasypkin and Tung-Ching Lee, (1998). Weights of the components mixed were calculated using the following formula:

$$C_{CF} = \frac{[r_{CF} \times M \times (100 - w)]}{[100 \times (100 - w_{CF})]} \tag{1}$$

$$C_{MF} = \frac{[r_{MF} \times M \times (100 - w)]}{[100 \times (100 - w_{MF})]} \tag{2}$$

$$W_X = M - C_{CF} - C_{MF} \tag{3}$$

C_{CF} and C_{MF} are the masses of cowpea flours (C_F) and millet flour (M_F), respectively, r_{CF} or r_{MF} are respective percentages of either cowpea flours (C_F) or millet flour (M_F) in the blend, d.b.; (r_{CF} + r_{MF} = 100%); M is the total mass of the blend; w, the moisture content of the final blend, percentage wet weight basis (w.w.b.); W_X is the weight of water added; and w_{CF} and w_{MF} are the moisture contents of C_F and M_F, respectively. The blends were mixed in a plastic bowl with the addition of the spices (Kimba & Ginger) at 1% level based on traditional formulation; and the whole mass packed in polyethylene bags which was kept in the refrigerator overnight to allow moisture equilibration. The samples were however brought to room temperature before extrusion process.

2.5 Experimental Design and Data Analysis

A three-factor three levels central composite rotatable composite design (CCRD) (Box and Hunter, 1957) was adopted to study the effect of feed composition (X_1), feed moisture content (X_2) and screw speed (X_3) on the expansion ratio, bulk density, proximate composition and amino acid profile during extrusion of pearl millet and cowpea flour mixtures for fura production. The outline of the experimental design is outlined in (Table 1).

Table 1. Independent Variables and Levels used for Central Composite Rotatable Design[1]

Variable	Symbol (X_i)	Coded variable level (x_i)				
		-1.68($\propto$)	-1	0	1	1.68($\propto$)
Feed composition (%)	X_1	3.2	10	20	30	36.8
Feed moisture (%)	X_2	16.6	20	25	30	33.4
Screw speed (Rpm)	X_3	116	150	200	250	284

[1]Transformation of coded variable (x_i) levels to uncoded variables (X_i) levels could be obtained from
$X_1 = 10x_1 + 20$; $X_2 = 5x_2 + 25$; $X_3 = 50x_3 + 200$

The levels of each variables were established according to literature information and preliminary trials. The outline of the experimental layout with the coded and natural values are presented in (Table 2). Homogeneous variances or homoscedasticity is a necessary prerequisite for (linear) regression models. Therefore, a reduction in variability within the objective response (dependent variables) was carried out by transforming the data to standardized scores ($z = \dfrac{x - \bar{x}}{s}$ where x = dependent variable of interest; $\bar{x}$ = mean of dependent variable of interest and s = standard deviation). For each standardized scores, analysis of variance (ANOVA) was conducted to determine significant differences among the treatment combinations. Also, data were analyzed using multiple regression procedures (SPSS, 2008). A quadratic polynomial regression model was assumed for predicting individual responses (Wanasundara and Shahidi, 1996). The model proposed for each response of Y was:

$$Y = b_0 + b_1 X_1 + b_2 X_2 + b_3 X_3 + b_{11} X_1^2 + b_{22} X_2^2 + b_{33} X_3^2 + b_{12} X_1 X_2 + b_{13} X_1 X_3 + b_{23} X_2 X_3$$
(4)

where Y = the response X_1 = Feed Composition, X_2 = Feed Moisture, X_3 = Screw Speed, b_0 = intercepts, b_1, b_2, b_3 are linear, b_{11}, b_{22}, b_{33} are quadratic and b_{12}, b_{13} and b_{23} are interaction regression coefficient terms. Coefficients of determination (R^2) were computed. The adequacy of the models was tested by separating the residual sum of squares into pure error and lack of fit. For each response, response surface plots were produced from the fitted quadratic equations, by holding the variable with the least effect on the response equal to a constant value, and changing the other two variables.

Table 2. Experimental design of extrusion experiment in their coded form and natural units[1,2]

Design point	Independent variables in coded form			Experimental variables in their natural units		
	(X_1)	(X_2)	(X_3)	(X_1)	(X_2)	(X_3)
1.	-1	-1	-1	10	20	150
2.	-1	+1	-1	10	30	150
3.	-1	-1	+1	10	20	250
4.	-1	+1	+1	10	30	250
5.	+1	-1	-1	30	20	150
6.	+1	+1	-1	30	30	150
7.	+1	-1	+1	30	20	250
8.	+1	+1	+1	30	30	250
9.	-1.68	0	0	3.2	25	200
10.	+1.68	0	0	36.8	25	200
11.	0	-1.68	0	20	16.6	200
12.	0	+1.68	0	20	33.4	200
13.	0	0	-1.68	20	25	116
14.	0	0	+1.68	20	25	284
15.	0	0	0	20	25	200

[1]Duplicate tests at all design point except the centre point (0, 0, 0) which was carried out five times and the result averaged. [2]Experiment was carried out in randomized order; (X_1) = feed composition (%), (X_2) = feed moisture (%) and (X_3) = Screw speed (rpm)

2.6 Extrusion Exercise

Extrusion cooking was performed in a single screw extruder, model (Brabender Duisburg DCE-330, Germany) equipped with a variable speed D-C drive unit, and strain gauge type torque meter. The screw has a linearly tapered rod and 20 equidistantly positioned flights. The extruder was fed manually through a screw operated conical hopper at a speed of 30 rpm which ensures the flights of the screw filled and avoiding accumulation of the material in the hopper. This type of feeding provides the close to maximal flow rate for the selected process parameters (constant temperature, constant die and screw geometry but with three variable screw speeds) and three designed feed composition and feed moisture contents. A round channel die with separate infolding heater was used. The die used was a cone shaped channel with 45 degrees entrance angle, a 3 mm diameter opening and 90 mm length. The screw was a 3:1 compression ratio. The inner barrel is provided with a grooved surface to ensure zero slip at the wall. The barrel is divided into two independent electrically heated zones that is (feed end and central zone). There is a third zone at the die barrel, electrically heated but not air cooled. The extruder barrel has a 20 mm diameter with length to diameter ratio (L:D) of 20:1. Desired barrel temperature was maintained by a circulating tap water controlled by inbuilt thermostat and a temperature control unit. The feed material was fed into a hopper mounted vertically above the end of the extruder which is equipped with a screw rotated at variable speed. The rotating hopper screw kept feed zone completely filled to achieve a 'choke fed' condition. Experimental samples were collected when steady state was achieved that is when the torque variation of ±0.28 joules (Nm) or about (0.5%) of full scale (Likmani et al.,1991). The extrusion process consisted of 15 individual runs and was conducted randomly.

2.7 Expansion Ratio (puff ratio)

Expansion ratio can be of two indices, diametral and longitudinal as described by Sopade and Le Grys (1991). Diametral expansion is defined as the diameter of the extrudate whilst longitudinal expansion is defined as the length per unit dry weight. The diameter was determined after cooling of the extrudate, 10 samples were assessed for each extrudate and for each sample; diameters at three different positions were taken using vernier calipers and the result averaged. Expansion ratio was expressed as the diameter of the extrudate to the diameter of the die.

2.8 Bulk Density of Extrudates

The bulk density (ρ) of extrudates was calculated using the methods described by Qing-Bo et al. (2005) as follows:

$$\text{Density } (\rho) = (4 \times \dot{m})/ (\pi \times D^2 \times L) \tag{5}$$

where $\dot{m}$ is the mass of extrudate with length L and diameter D. The samples were randomly selected and replicated 10 times and the average result value taken.

2.9 Moisture Analysis

Moisture contents of raw and extrudates were determined as described by (AOAC, 1984). Triplicate determinations were carried out and the result averaged.

2.10 Crude Fat Determination

Crude fat of samples was determined using soxhlet fat extraction system (AOAC, 1984).

2.11 Crude Protein Determination

Protein content was determined by Kjeldahl Method (AOAC, 1984). Triplicate determinations were carried out and the result averaged.

2.12 Ash Determination

Ash was determined by the method of AOAC (1984). Triplicate determinations were carried out and the result averaged.

2.13 Determination of Carbohydrates

The percentage carbohydrate was determined by difference (Egan et al., 1981).

2.14 Amino Acid Analysis

The method of Sotelo et al. (1994) was used in determining the amino acid content of the extrudates. One gram of sample was dissolved in 20 ml of 6N HCL. This was then poured into a hydrolysis tube with screw cap and hydrolyzed for 22 hour under a nitrogen atmosphere. The acid was evaporated using a rotary evaporator and residue washed three times with distilled water. The extracted sample was dissolved in 1ml acetate buffer of pH 3.1. After dilution to a known volume, the hydrolysate was transferred into a Beckman system (model 6300) high performance amino acid analyzer. Amino acid scores were calculated as gram per 100 gram protein (g/100g protein). Triplicate determinations were carried out and the result averaged.

2.15 Sensory Evaluation

One hundred gram of pulverized extruded fura was added to 500 ml of water in a 1000ml beaker each for all the 15 extruded fura samples, 100g sugar was added to each of the preparations based on traditional recipe and these samples were used for sensory evaluation test. Twenty untrained member judges (28 – 45 year old males and females) were employed. The prepared samples were placed on tables inside a plastic bowl provided with individual booths. The panelists were to taste and swallow each of the fura gruel and rinse their mouth with tap water between samples. The panelists were familiar with fura and were asked to indicate their opinion on four sensory attributes, namely the colour, flavour, texture (gritness) and the overall acceptability using Hedonic scale rating of 9 = liked extremely to 1 = disliked extremely. Acceptability scores using the mean of the observations were recorded.

3. RESULTS AND DISCUSSIONS

3.1 Model Description

Studies was conducted using the response surface method in modeling the chemical composition and amino acid profile of extruded millet – cowpea based fura as affected by the process variables feed composition level of cowpea added to millet (X_1), feed moisture (X_2), and screw speed (X_3). The independent and dependent variables were fitted to the second – order model equation and examined for the goodness of fit. The analysis of variance were performed to evaluate the lack of fit and the significance of the linear, quadratic and interaction effects of the independent variables on the dependent variables. The lack of fit test is a measure of the failure of a model to represent data in the experimental domain at which points were not included in the regression (Varnalis et al., 2004). Coefficient of determinant, R^2, is defined as the ratio of the explained variation to the total variation and is a measure of the degree of fit (Singh et al., 2007). It is also the proportion of the variability in the response variables, which is accounted for by the regression analysis (Mclaren et al., 1977). When R^2 approaches unity, the better the empirical model fits the actual data. The smaller is R^2, the less relevant the dependent variables in the model have in explaining the behavior of variation (Myers and Montgomery, 2002). It is suggested that for good fit model, R^2 should be at least 80%. The results showed that the model for all the response variables were highly adequate because they have satisfactory levels of R^2 of more than 80% and that there is no significant lack of fit in all the response variables. If a model has a significant lack of fit, it is not a good indicator of the response and should not be used for prediction (Myers and Montgomery, 2002). We may probably conclude that the proposed models approximates the response surfaces and can be used suitably for prediction at any values of the parameters within experimental range.

3.2 Expansion Ratio (ER) and Bulk Density (BD)

The expansion characteristic of extruded snacks is important especially for the acceptability by the consumer. Most extruded snack products are expected to have a puffed structure. Property index such as expansion ratio and density may be used to quantify this structure. Sectional expansion is a measure of sectional expansion area of fura extruded. The mean values of expansion ratios (ER) and bulk density (BD) of the extruded millet – cowpea fura are shown in (Table 3). From the result it shows that lowest value for the ER was 1.6 and the highest value was 4.3 representing (20% cowpea, 33.4 % feed moisture and 200 rpm screw speed) and (20% cowpea, 25 % feed moisture and 284 rpm screw speed) respectively. It appears that ER decreased with increasing moisture content. The expansion ratio is a vital factor for instant products because of the short hydration time for which this study was intended. Bulk density has been linked with the expansion ratio in describing the degree of puffing in extrudates. The observed values for the bulk density (BD) varied from 0.1 – 0.4 kgm^{-3} for both samples (10% cowpea, 20 % feed moisture and 250 rpm screw speed; 20% cowpea, 25 % feed moisture and 284 rpm screw speed) and (20% cowpea, 33.4 % feed moisture and 200 rpm screw speed) respectively. Response surfaces analysis was applied to the experimental data and a quadratic polynomial regression response surface model Eq. (4) was fitted to all the response parameters. The multiple regression equation representing the effect of processing variables on ER and BD is given by the second order model; the models explained the variability in the experimental data obtained. The sign and magnitude of coefficients in equations (6) and (7) indicate the effect of parameters on the response. A negative sign of coefficient indicates a decrease in the response when the level of the parameter is increased, while a positive sign indicates increase in the response with increase in the level of the parameter:

$$ER = 0.25 + 0.01X_1 - 0.72X_2 + 0.56X_3 - 0.05X_1^2 - 0.13X_2^2 + 0.54X_3^2 + 0.11X_1X_2 - 0.23X_1X_3 - 0.23X_2X_3$$
$$(6)$$

$$BD = 0.19 + 0.09X_1 + 0.87X_2 - 0.53X_3 - 0.07X_1^2 + 0.21X_2^2 - 0.42X_3^2 + 0.08X_1X_2 + 0.07X_1X_3 + 0.03X_2X_3$$
$$(7)$$

where X_1 refers to coded value of feed composition, X_2 refers to the coded value of feed moisture content, X_3 refers to coded value of screw speed.

Analysis of variance (table not shown) indicates that the linear effects of the independent variables (feed moisture content and screw speed) significantly ($P<0.05$) affected the expansion ratio (Table 3). The model of the regression equation showed good fit with $R^2 = 0.92$ for ER. This suggests a very good fit to the experimental data and the model could be used to describe the process. The negative significant ($P<0.05$) effect of linear coefficients of feed moisture indicated that this variable influenced the ER in reverse order. This observation is in agreement with Kokini et al. (1991) who reported that moisture plays a key role in the mechanism responsible for expansion. Harmann and Harper (1973) postulated two factors in governing expansion: (a) dough viscosity, and (b) elastic force (die swell in the extrudate. The elastic forces will be dominant at low moisture and temperature. The bubble growth, which is driven by the pressure difference between the interior of the growing bubble and atmospheric pressure resisted primarily by the viscosity of the bubble wall, will usually dominate the expansion at high moisture content and high temperature (Panmanabhan & Bhattacharyya, 1989). A viscoelastic melt in a food extruder expands due to flashing of moisture at the die exit. The expansion process can be described as nucleation in the die, extrudate swelling immediately beyond the die, followed by bubble growth and collapse (Kokini et al., 1991). Increased water content in the melt would soften the amylopectin molecular structure and reduce its elastic characteristics to decrease diametral expansion (Alvarez-Martinez et al., 1988). Feed moisture has been identified as the main factor affecting extrudate expansion and density (Gujral et al., 2001). Extrudates can expand in both the cross-sectional (diametrical) direction and the longitudinal direction (Launary and Lisch, 1983). A porous, expanded, sponge-like structure is formed inside extrudates as a result of many tiny steam bubbles created by the rapid release of pressure after exiting the die (Conway, 1971). Increased feed moisture during extrusion would provoke change in the amylopectin molecular structure of the material reducing the melt elasticity thus decreasing the expansion and increasing the density of extrudate (Qing-Bo et al., 2005). The extrusion variable screw speed indicated significant ($P<0.05$) effect on the ER positively. The quadratic effect of the variable screw speed significantly ($P<0.05$) influenced the ER. Seker, (2005) reported that increasing screw speed improved sectional expansion and reduced bulk density of extrudate during extrusion of soybean protein and corn starch. The extrusion variables feed composition i.e. level of cowpea and the feed moisture had negative quadratic effect on the ER but it was not significant ($P<0.05$). There was marginal increase in ER as cowpea flour was added. Though some workers have reported increase in levels of whole bean flour resulted in a significant decrease in expansion. They attributed the decrease as a result of interference of fibre with bubble formation, this is on the basis that fibre can rupture cell walls and prevent air bubbles from expanding to their maximum potential. The disagreement of our result with this report may be attributed to that fact that we used dehulled flour for our studies which resulted in reduced fibre content. Ainsworth et al. (2007) reported increased expansion of snack produced from mixtures of corn and dehulled chickpeas flours.

The BD regression equation coefficients for millet – cowpea based fura is presented in (Table 5). The expansion ratio (BD) was influenced significantly by linear and quadratic terms significantly ($p<0.05$). The interaction term effect did not affect BD significantly ($p>0.05$). The coefficient of determinant R^2 for the BD was 0.94, which shows good fit for the model in describing the effects of feed composition, feed moisture and screw speed on the BD. It was observed that an increase in screw speed resulted in an extrudate with lower density. Higher screw speeds may be expected to lower melt viscosity of the mix which can increase the elasticity of the dough, resulting in a reduced extrudate density. Increasing screw speed tends to increase the shearing effect, this causes protein molecules to be stretched farther apart, weakening bonds and resulting in a puffer product. Longer residence times associated with low screw speeds may be responsible for decreased ER, which suggests that such materials require more shear for a better developed dough, resulting in better expansion. Singh et al. (2007) reported decrease in ER with increase in feed moisture of rice – pea grits extrudates. Extruded snacks possess the typical texture of puffed, light and crispy. Some physical properties of extruded snack were reported including bulk density of 48-64 g/L, 50-160 g/L (Moore, 1994) and 59 10g/L (Boonyasirikool et al., 1996) and expansion ratio of 3.1 - 3.8 (Mohamed, 1990) and 4.03 (Boonyasirikool et al., 1996). These relationships have been reported elsewhere for corn and wheat based snacks (Ilo et al., 1999).

Table 3. Experimental design and observed values of expansion ratio (ER), bulk density (BD) and proximate composition of extruded millet-cowpea based fura

Independent variables[b]			Dependent variables[a]						
X_1	X_2	X_3	ER(Ratio)	BD(Kgm^{-3})	CHON	FAT	FAT	ASH	HOH
10	20	150	2.8 ± 0.7	0.2 ±0.03	12.7±0.3	3.2±0.3	3.2±0.3	1.8±0.3	5.4±0.3
10	30	150	1.7 ± 0.8	0.3 ± 0.05	12.9±0.5	3.2±0.3	3.2±0.3	1.8±0.1	5.2±0.2
10	20	250	4.2 ± 1.0	0.1 ±0.02	12.9±0.2	3.2±0.7	3.2±0.7	1.7±0.1	5.2±0.3
10	30	250	2.3 ± 0.4	0.2 ± 0.08	12.7±0.5	3.2±0.3	3.2±0.3	1.8±0.2	5.5±0.4
30	20	150	2.8 ± 0.3	0.2 ±0.06	15.8±0.5	1.7±0.2	1.7±0.2	1.9±0.2	5.0±0.3
30	30	150	2.2 ± 0.2	0.3 ±0.02	15.8±0.8	1.7±0.4	1.7±0.4	1.9±0.2	5.1±0.4
30	20	250	3.5 ± 0.3	0.1 ±0.03	15.7±0.4	1.7±0.4	1.7±0.4	2.0±0.2	5.7±0.7
30	30	250	2.0 ± 0.8	0.3 ±0.01	15.8±0.5	1.5±0.4	1.5±0.4	2.0±0.5	5.3±0.2
3.2	25	200	2.0 ± 0.7	0.3 ±0.01	11.2±0.3	3.6±0.3	3.6±0.3	1.7±0.2	5.7±0.4
36.8	25	200	2.3 ± 0.6	0.3 ±0.01	16.8±0.5	2.0±0.3	2.0±0.3	2.0±0.2	5.3±0.3
20	16.6	200	2.7 ± 0.7	0.2 ±0.01	14.5±0.7	2.9±0.3	2.9±0.3	1.8±0.3	5.3±0.5
20	33.4	200	1.6 ±0.5	0.4 ±0.02	14.3±0.3	2.9±0.7	2.9±0.7	1.8±0.3	5.6±0.3
20	25	116	2.4 ±0.7	0.3 ±0.02	14.3±0.8	3.0±0.4	3.0±0.4	1.9±0.2	5.3±0.7
20	25	284	4.3 ± 0.6	0.1 ±0.04	14.4±0.9	2.7±0.3	2.7±0.3	1.8±0.1	5.3±0.2
20	25	200	2.3 ± 0.9	0.2 ±0.03	14.3±0.5	2.9±0.3	2.9±0.3	1.9±0.3	5.5±0.2

bX1 = Feed composition (%); X2 = Feed moisture (%) and X3 = Screw speed (rpm). CHON = Protein; FAT = Fat; CHO = Carbohydrate; ASH = Ash; HOH = Water. aValues are means and ± standard deviation of triplicate determinations.

3.3 Proximate Composition

The mean result of proximate composition of fura extrudates is shown in (Table 3). From the result it shows that the mean observed value of protein for the fura extrudates ranged from 11.2 – 16.8 % suggesting proportional increase in protein content with fortification of pearl millet with cowpea flour. Protein content was measured as nitrogen (N x 6.25); hence the apparent protein content was not affected by extrusion temperature, as nitrogen is not affected by heat treatment (Pelembe et al., 2002). The average value of the fat content for the fura extrudates showed a general low values which ranged from 1.5 – 3.6 %. The mean observed values for the carbohydrates relatively remained within a narrow range of 74.1 – 77.3 %. The carbohydrate level relatively remained high as expected as cowpea is equally high in carbohydrates. The ash content showed that it remained within a narrow range of 1.7 – 2.0 %. Regression coefficients for objective responses for extruded fura composition are presented in (Table 5). Analysis of variance (table not shown) indicates that linear and quadratic effects significantly ($P<0.05$) affected the protein content of fura extrudates as expected. The result indicated that increasing the amount of cowpea flour resulted in linear increase in the protein contents of extruded fura. The influence of feed moisture and screw speed independent variables linear terms indicated negative effects on the protein content but was not significant ($P<0.05$). This negative effect suggests that increasing the feed moisture and screw speed resulted in decreased amount of protein content in the extrudate. This reduction in protein content may be attributed to the complex nature of interactions between extruder conditions, these changes might not be related to a single factor. Hence, the role of feed moisture, screw speed and other interactions of other parameters on the protein nutritional value is a point that obviously needs further investigation. The quadratic effect of feed composition showed significant ($P<0.05$) influence on the protein content. The effect of square term on the protein content was a positive effect. However, there was no significant effect observed by the interaction terms on the protein content of extrudates. The regression analysis for the carbohydrate indicated that the linear effect of feed composition affected the carbohydrate contents significantly ($P<0.05$). This result indicates that increase in the amount cowpea flour translates to lower values of carbohydrates of the fura extrudates. The linear effect of feed moisture and screw speed indicated insignificant ($P<0.05$) effect on the carbohydrate content of extrudates. Experimental design points 5, 6, 7, 8, & 10 have attained the least protein contents if compared with the FAO/WHO/UNU (1985) minimum protein level of 15.7 % recommended for supplementary mixtures of protein (Table 3).

The carbohydrate and fat content was significantly ($P<0.05$) influenced by the linear term. This influence was negative effect suggesting that increase in the amount of cowpea flour resulted in the decrease in the carbohydrates and fat contents of fura extrudates. This effect is expected as millet flour has higher amounts of carbohydrates and fats than cowpea (result not shown). It appears that increasing the level of cowpea flour resulted in decreased fat content; as both millet and cowpea are low in fat content. In addition it may be attributed to the fact that cowpea is relatively lower in fat than millet (result not shown). An implication of this low fat content is the need for fat fortification in millet – cowpea mixtures, to meet up with the minimum fat requirement of 6% for complementary formulations (Mitzner et al., 1984). However, the low energy content as a result of low fat can be compensated for by the reduced viscosity due extrusion, therefore allowing a high level of solid matter in the gruel. The linear effect of feed composition indicated significant ($P<0.05$) influence on the ash content. The effect was a positive, suggesting increase in the ash content as cowpea flour was added to millet flour. Though generally the amount of ash was so low in the formulations of this study, Obatolu (2002) similarly observed low ash contents of mixtures of cereal/ legume blends. This could be explained probably because cowpea flour relatively has more amount of ash content when compared with millet flour (result not shown).

The regression models fitted to the experimental data showed high coefficients of determinants with R^2 = 0.96, 0.94, 0.94, 0.85 and 0.80 for protein (CHON), carbohydrate (CHO), fat (FAT), ash (ASH) and water (HOH) respectively. This indicates that the regression models can be considered adequate to study response tendencies. Among the three extrusion variables feed composition was analyzed to be most important factor affecting the proximate composition of fura extrudates.

The major constituents of pearl millet flour are carbohydrate (70.8%) and protein (10.8%) (result not shown). This study has shown increase in the protein content of millet as a result of fortification with cowpea flour. In addition legume proteins are rich sources of lysine and threonine (Alonso et al., 2000a). From the view point of protein content, the basis of advocating cereal – based complementation as a method of improving the protein content of cereal – based traditional foods is justified by the results of this study. Several workers have also reported marked improvements in the protein content of cereals when fortified with legumes (Lazou and Krokida, 2010).

3.4 Amino Acid Profile

The mean result of amino acid profile of the fura extrudates is presented in (Table 4). As already known, lysine and methionine are generally limiting in cereals and legumes, respectively. From the result it shows that the mean observed value of lysine for the extrudates ranged from 5.1 - 6.6g /100g protein; representing (feed composition 3.2% cowpea, 25% feed moisture content and 200rpm screw speed) and (36.8% cowpea, 25% feed moisture content and 200rpm screw speed) respectively. The methionine content ranged from 1.3 - 3.8g /100g protein; representing (36.8% 25% and 200rpm screw speed) and (feed composition 3.2% cowpea, 25% feed moisture content and 200rpm screw speed) respectively (Table 4). Response surfaces analysis was applied to the experimental data and a quadratic polynomial regression response surface model Eq. (4) was fitted to all the response parameters. The multiple regression equation representing the effect of processing variables on amino acid profile is given by the second order model; the models explained the variability in the experimental data obtained. The sign and magnitude of coefficients in equations (8, 9, 10, 11, 12, 13 and 14) for the amino acids content of fura samples indicating the effect of process variables on their responses. A negative sign of coefficient indicates a decrease in the response when the level of the parameter is increased, while a positive sign indicates increase in the response with increase in the level of the parameter:

$$LYSINE = 0.15 - 0.67X_1 - 0.0.12X_2 - 0.22X_3 - 0.47X_1^2 + 0.02X_2^2 + 0.26X_3^2 - 0.13X_1X_2 + 0.2X_1X_3 + 0.33X_2X_3 \qquad (8)$$

$$LEUCINE = 0.54 - 0.70X_1 + 0.08X_2 - 0.46X_3 - 0.45X_1^2 - 0.24X_2^2 + 0.01X_3^2 - 0.22X_1X_2 + 0.10X_1X_3 + 0.38X_2X_3 \qquad (9)$$

$$VALINE = -0.49 + 0.91X_1 + 0.03X_2 + 0.004X_3 + 0.34X_1^2 + 0.4X_2^2 - 0.12X_3^2 - 0.02X_1X_2 - 0.04X_1X_3 - 0.8X_2X_3 \qquad (10)$$

$$METHCYT = -0.96 + 0.67X_1 - 0.01X_2 + 0.03X_3 + 0.75X_1^2 + 0.24X_2^2 + 0.21X_3^2 + 0.02X_1X_2 + 0.05X_1X_3 - 0.02X_2X_3 \qquad (11)$$

$$THRENINE = 0.56 + 0.85X_1 + 0.004X_2 + 0.01X_3 - 0.56X_1^2 - 0.06X_2^2 - 0.08X_3^2 - 0.02X_1X_2 + 0.01X_1X_3 + 0.01X_2X_3 \qquad (12)$$

$$TRYPHYL = -0.15 + 1.03X_1 + 0.01X_2 - 0.02X_3 + 0.16X_1^2 - 0.02X_2^2 + 0.05X_3^2 + 0.02X_1X_2 + 0.004X_1X_3 - 0.04X_2X_3 \qquad (13)$$

$$TRYTPHAN = -0.31 - 0.95X_1 - 0.08X_2 - 0.041X_3 + 0.24X_1^2 + 0.13X_2^2 + 0.01X_3^2 - 0.13X_1X_2 - 0.02X_1X_3 + 0.05X_2X_3 \qquad (14)$$

where X_1 refers to coded value of feed composition, X_2 refers to the coded value of feed moisture content, X_3 refers to coded value of screw speed.

Regression coefficients for objective responses for fura extrudates amino acid profile are presented in (Table 5). Analysis of variance (table not shown) indicates that the amino acid content were influenced significantly ($P<0.05$) by the linear and quadratic terms for lysine, leucine, valine and threonine. The linear effects of feed composition was positive for lysine content, which suggests increasing the amount of cowpea resulted in increased in the amount of lysine content of extrudates. The lysine content was however negatively influenced by the linear effects of feed moisture and screw speed. This result confirms the same effects of feed moisture and screw speed on protein content earlier mentioned. The quadratic effect of feed composition influenced the lysine content of extrudates significantly ($P<0.05$). The result indicates positive effect of quadratic term on the lysine contents of extrudates. The i/leucine was influenced significantly ($P<0.05$) by the quadratic term effect of feed composition. The result indicates positive effect by the quadratic effect of feed composition on the i/leucine content. The amino acid leucine content of fura extrudates was influenced significantly ($P<0.05$) by the linear term of feed composition and screw speed. This effect was negative which suggested that there was decrease in the leucine content of extrudates as cowpea flour and screw speed was increased.

The coefficients of determinant $R^2 =$ 0.90, 0.85, 0.86, 0.92, 0.88, 0.85 and 0.93 for lysine, i/leucine, leucine, valine, methionine – cystine, threonine and tryptophan respectively, suggesting good fit of the model (Table 5). From the result of this study it shows that the level of lysine contents of fura extrudates increased generally as the amount of cowpea flour was increased. Tyrosine-phenylalanine, i/leucine, valine and leucine all increased as the level of cowpea flour was increased in the extrudates, from the regression analysis Table 6. Similar observations have been reported by other authors for blends of cereals and legumes (Obatolu, 2002). Nkama and Malleshi (1998) reported that, lysine increased by 75% as a result of supplementation of millet with cowpea at (83:17) ratio. They also reported similar increase of other essential amino acids as a result of supplementation; these amino acids include histidine, threonine, valine and i/leucine. Legumes provide a larger protein intake and amino acid balance when consumed with cereals, which significantly improves the protein quality (Bressani, 1975). Pelembe et al. (2002) reported that protein content of extrudates increased proportionally with the amount of cowpea flour in sorghum. The mean values of the amino acids profile from this study revealed that some of the essential amino acids were present in adequate amount if compared with the recommended values of FAO/WHO (1973). From the point of view of utilization, cooking quality of grain legumes is very important. It may be expected that grain legumes will be utilized more extensively if quick legume processing technology is adopted on a commercial basis and more acceptable, nutritious and digestible food products are developed from such technologies like extrusion cooking. It can be inferred that extrusion cooking and supplementation of pearl millet and cowpea flour mixtures can be employed in the improvement of the nutritional quality of traditional or millet based foods like fura.

Digestibility is considered the most determinant of protein quality in adults, according to FAO/WHO/UNU (1985). Although digestibility test was not conducted in this study, Singh et al. (2007) reported that protein digestibility value of extrudates is higher than non extruded products. The possible cause might be denaturation of proteins and inactivation of antinutritional factors that impair digestion especially in cowpea which was used as a basic ingredient in this study. The nutritional value of vegetable protein is usually enhanced by mild extrusion cooking conditions, owing to increase in digestibility (Areas, 1992). Benefits of beans extrusion – cooking are deactivation of heat labile inhibitors (Aguilera, et al., 1984). An advantage of destruction is the destruction of antinutritional factors, especially trypsin inhibitors, haemagglutinins, tannins and phytates, all of which inhibit protein digestibility (Singh et al., 2007). Extrusion has been shown to be very effective in reducing or eliminating lectin activity in legume flour (Alonso et al., 2000ab).

Thus, extrusion cooking is more effective in reducing or eliminating lectin activity as compared with traditional aqueous heat treatment.

3.5 Extrudate Photographic Responses

The visual effects of extrusion variables (feed composition, feed moisture and screw speed) can be seen as shown in Plates 1-15. The effect of the independent variables on the expansion ratio and colour of extrudates is evident. The results shown in the photographs describes the changes occurred during extrusion as influenced by the extrusion variables. It is evident that design points 3, 4 and 12 had the least expansion because of the high feed moisture levels 30, 30 and 33.4% respectively. These samples consequently have smooth body and darker colour because they are denser. Those samples with higher expansion ratios indicated more wrinkles and lighter colour as expected for design points 1, 3, 5, 7, 10, and 11 as a result of fading of colour with increase in expansion resulting from the air bubbles produced during extrusion. Colour changes in extruded products have been reported to be due to decomposition of pigments, product expansion causing colour fading and chemical reactions such as caramelisation of carbohydrates (Chen et al., 1991).

3.6 Sensory Evaluation

The statistical sensory qualities of millet – cowpea fura extrudates is presented in (Table 6). From the result it shows that design point 11 representing 20 % feed composition (X_1), 16.6 % feed moisture (X_2) and 200 rpm screw speed (X_3) recorded the highest value of acceptability for colour. This design point was significantly different (P<0.05) from other extrudates with the exception of design point 13 representing 20 % feed composition, 25 % feed moisture and 116 rpm screw speed. The flavour shows that there was no significant difference (P<0.05) between design point 1, 3, 5, and 11 (Table 6). The same design point 11 recorded the same pattern of highest acceptance for the flavor, texture and overall acceptability of extrudates. The design point 11 had the lowest processing extrusion feed moisture and recorded the highest sectional expansion and lowest bulk density (result not shown) probably influenced the acceptability of this sample more than others. Due to more expansion which resulted in fading of the dark colour of extrudate to brighter form might have made the sample much acceptable to the panelists. Colour changes in extruded products have been reported to be due to decomposition of pigments, product expansion causing colour fading and chemical reactions such as caramelisation of carbohydrates (Chen et al., 1991).

3.7 Optimization

Numerical optimization was carried out for the process variables for processing of fura extrudates from millet – cowpea mixtures for obtaining the best product. Desired goals were assigned for all the parameters for obtaining the numerical optimization values for the responses. All the processing parameters were kept in range. Lysine, protein content and expansion ratio were maximized while bulk density was minimized. The optimum values obtained for feed composition, feed moisture and screw speed are 36.5% cowpea level, and 22.3% feed moisture and 186.7 rpm respectively. The corresponding optimum values of lysine, protein content, expansion ratio and bulk density are 6.6g/100g protein, 16.6%, 2.80 and 0.52 Kg m^{-3} respectively.

Table 4. Experimental design and observed amino acid profile of millet – cowpea based fura extrudates (g/100 g protein)

Independent variables[b]			Dependent variables[a]							
X_1	X_2	X_3	[1]Lysine	[2]I/Leucine	[3]Leucine	[4]Valine	[5]Meth+Cyt	[6]Threonine	[7]Tyr+Phyl	[8]Tryptophan
10	20	150	5.2±0.1	2.9±0.1	9.1±0.3	3.2±0.3	3.5±0.1	3.6±0.3	4.3±0.3	0.9 ±0.01
10	30	150	5.2±0.3	2.8±0.2	9.2±0.6	3.2±0.5	3.5±0.1	3.8 ±0.3	4.2±0.4	0.9 ±0.04
10	20	250	5.2±0.5	2.8±0.2	9.2 ±0.8	3.3±0.3	3.5±0.2	3.8±0.2	4.3±0.4	0.9±0.02
10	30	250	5.2±0.3	2.8±0.2	9.1±0.4	3.1±0.2	3.5±0.2	2.7±0.2	4.2±0.3	0.8±0.01
30	20	150	5.7±0.6	3.1±0.4	10.3±0.7	4.3±0.5	1.9±0.1	4.2±0.3	4.8±0.2	1.0±0.05
30	30	150	5.9±0.4	3.1±0.2	10.1±0.6	4.3±0.3	1.7 ±0.2	4.2±0.2	4.8±0.4	1.0±0.03
30	20	250	5.8±0.5	3.2±0.5	10.2±0.5	4.4±0.6	1.8±0.2	3.3±0.3	5.0±0.5	1.0±0.08
30	30	250	5.8±0.3	3.2±0.4	10.3±0.5	4.3±0.8	1.8 ±0.2	3.2±0.3	4.8±0.3	1.0±0.06
3.2	25	200	5.1±0.5	2.5±0.2	6.9±0.4	2.4±0.4	3.8±0.2	3.4±0.2	5.9±0.3	0.7±0.02
36.8	25	200	6.6±0.8	3.4±0.3	11.4±0.5	5.5±0.5	1.3±0.3	4.6±0.2	4.7±0.3	1.3±0.05
20	16.6	200	5.9±0.4	2.5±0.4	10.5±0.7	3.5±0.7	2.0±0.2	4.4±0.3	4.5±0.2	1.2±0.04
20	33.4	200	6.0±0.7	2.6±0.4	10.6±0.4	3.7±0.5	2.0±0.2	4.3±0.3	4.5±0.2	1.3±0.06
20	25	116	5.2±0.4	2.5±0.3	10.5±0.4	3.8±0.4	1.9±0.2	4.4±0.3	4.5±0.2	1.2±0.04
20	25	284	5.4±0.8	2.5±0.4	10.6±0.6	3.6±0.4	1.8±0.2	4.4 ±0.2	4.5±0.3	1.3±0.02
20	25	200	5.4±0.7	2.5±0.2	10.5±0.6	3.5±0.4	1.8±0.2	4.3±0.4	4.44±0.3	1.2±0.07

[b]X_1 = Feed composition (%); X_2 = Feed moisture (%) and X_3 = Screw speed (rpm). [a]Values are means and ± standard deviation of triplicate determinations. 1=Lysine; 2= Isoleucine; 3= Leucine; 4=Valine; 5=Methionine + Cystine; 6=Threonine; 7= Tyrosine + Phenylalanine; 8 =Tryptophan

Table 5. Regression equation coefficients for objective responses a, b expansion ratio (ER), bulk density (BD), proximate composition and amino acid

Coefficient	ER	BD	[1]CHON	[2]CHO	[3]FAT	[4]ASH	[5]HOH	[6]LYSINE	[7]I/LECNE	[8]LEUCINE	[9]VALINE	[10]MET-CYT	[11]TRENE	[12]TRPTPHN
Linear
B_0	-0.246	0.188	0.676*	-0.169	-0.306	0.491	-0.161	0.144	-0.844	0.539	-0.492	-0.959	0.558	-0.152
B_1	0.014	0.088	1.289*	0.107*	1.254*	0.772*	-0.535	0.665*	0.650	0.691*	0.908*	0.673	0.849*	1.030*
B_2	-0.722*	0.866*	-0.158	-1.6 5	-0.133	0.038	0.471	-0.123	0.103	0.080	0.035	-0.011	0.004	0.013
B_3	0.586*	0.527*	-0.264	-1.404	0.364	-0.067	0.589	-0.220	0.065	0.458*	0.004	0.027	0.012	-0.027
Quadratic
B_{11}	-0.047	-0.067	0.108*	0.621	0.166	-0.324	0.179	0.468*	0.729*	0.445*	0.336*	0.745*	-0.562*	0.164
B_{22}	-0.133	0.210*	-0.285	0.690	-0.214	-0.373	-0.488	0.024	0.305	-0.239	0.396*	0.242	-0.057	-0.021
B_{33}	0.541*	0.419*	-0.346	0.012	-0.217	-0.022	0.289	0.265	0.142	0.012	-0.119	0.208	-0.076	0.0467
Interaction
B_{12}	0.112	0.076	-0.249	0.529	0.343	0.172	-0.723	-0.127	-0.259	-0.219	-0.018	0.020	-0.021	0.018
B_{13}	-0.219	0.068	-0.134	-0.470	-0.224	0.557*	0.260	0.198	-0.324	0.101	-0.036	0.053	0.006	0.004
B_{23}	0.282*	0.079	-0.189	0.727	-0.173	-0.089	-0.316	0.333	0.266	0.377	-0.079	-0.020	0.013	-0.037
R^2	0.933	0.935	0.963	0.939	0.938	0.849	0.801	0.897	0.848	0.855	0.921	0.879	0.849	0.929
Adjusted R^2	0.873	0.869	0.882	0.836	0.938	0.623	0.343	0.836	0.894	0.880	0.821	0.874	0.856	0.838
Lack of fit	*	NS	NS	NS	NS	NS	NS	NS	NS	NS	NS	NS	NS	NS
Model	*	*	*	*	*	*	*	**	*	*	*	*	*	*

[a] $Y = b_0 + b_1 X_1 + b_2 X_2 + b_3 X_3 + b_{11} X_1^2 + b_{22} X_2^2 + b_{33} X_3^2 + b_{12} X_1 X_2 + b_{13} X_1 X_3 + b_{23} X_2 X_3$ $X_1 =$ Feed Composition, $X_2 =$ Feed Moisture, $X_3 =$ Screw Speed[b]*, ** Significant at $P < 0.05$ and $P < 0.01$, respectively; NS, not significant.

1=Protein2=Carbohydrate3=Fat4=Ash5=Water6=Lysine7=Isoleucine8= Leucine9=Valine10=Methionine-Cystine11=T hreonine12=Tryptophan

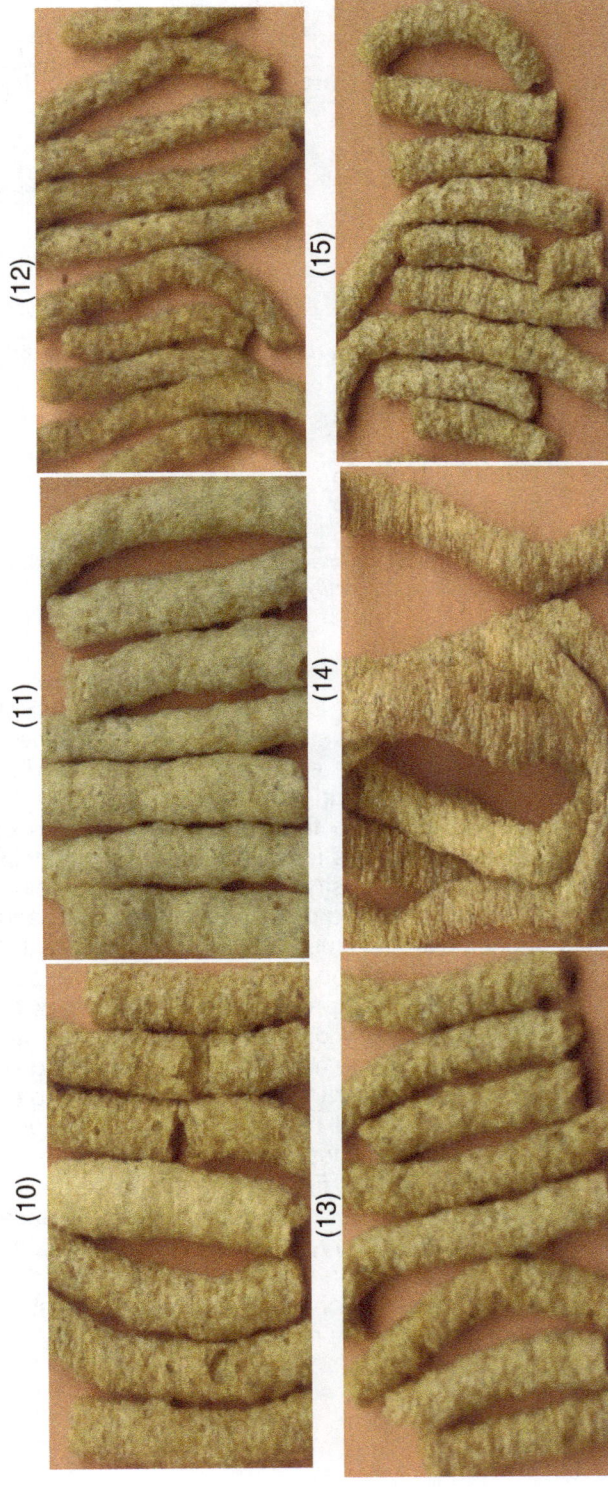

Plates 1 – 15: (1) 10% Cowpea, 20% moisture, 150rpm; (2) 10% Cowpea, 30% moisture, 150rpm; (3) 10% Cowpea, 20% moisture, 250 rpm; (4) 10% Cowpea, 30% moisture, 250rpm; (5) 30% Cowpea, 20% moisture, 150rpm; (6) 30% Cowpea, 30% moisture, 150rpm; (7) 30% Cowpea, 20% moisture, 250rpm; (8) 30% Cowpea, 30% moisture, 250rpm; (9) 3.2% Cowpea, 25% moisture, 200 rpm; (10) 36.8% Cowpea, 25% moisture, 200rpm; (11) 20% Cowpea, 16.6% moisture, 200rpm; (12) 20% Cowpea, 33.4% moisture, 200rpm; (13) 20% Cowpea, 25% moisture, 116rpm; (14) 20% Cowpea, 25% moisture, 284rpm; (15) 20% Cowpea, 25% moisture, 200rpm;

Table 6. Sensory qualities of extruded millet – cowpea mixtures[a]

Independent variables[b]			Dependent variables[a]			
X_1	X_2	X_3	Colour	Flavour	Texture	Overall/Acceptability
10	20	150	6.45b	6.87ab	6.67b	6.50b
10	30	150	6.30b	6.70b	6.43b	5.34c
10	20	250	6.54b	6.88ab	6.88ab	6.66b
10	30	250	6.30b	6.34b	6.38	6.45b
30	20	150	6.41b	6.87ab	6.56b	6.55b
30	30	150	5.50c	6.12b	6.34b	5.59c
30	20	250	5.40c	6.34b	6.41b	5.65c
30	30	250	5.60c	5.55c	5.65c	5.78c
3.2	25	200	6.35b	6.78b	6.57b	6.44b
36.8	25	200	5.36c	5.56c	5.65c	5.34c
20	16.6	200	7.30a	7.56a	7.54a	7.55a
20	33.4	200	5.89bc	5.76c	5.79c	5.68c
20	25	116	6.89ab	6.76b	6.66b	6.75b
20	25	284	5.53c	5.25c	6.54b	5.55c
20	25	200	5.40c	5.35c	6.53b	5.34c

[b]X_1 = Feed composition (%); X_2 = Feed moisture (%) and X_3 = Screw speed (rpm). [a]Mean values in the same column with different letters are significantly different (P<0.05)

4. CONCLUSION

Designed experiments were conducted following Response Surface Methodology (RSM) for the extrusion of pearl millet and cowpea based fura using a single screw extruder. The RSM was found to be effective technique to investigate the expansion ratio, bulk density, proximate composition and amino acid profile of fura extrudates. From the result it shows that the three extrusion variables were found to influence the extrudate properties either independently or interreactively. Feed moisture content was found to be the most significant factor that affected the expansion ratio and the bulk density, The extrusion variable feed composition i.e the level cowpea was found to influence the proximate composition and the amino acid profile. The high correlation coefficient of multiple determinations at 95% confidence level and the model equation developed can be used for predicting expansion ratio, bulk density, proximate composition and amino acid profile of millet – cowpea based fura. Using RSM the combined effect of three variables on extrudate response can be predicted which is difficult to achieve with conventional methods. The understanding of the effect of the process variables on quality parameters of fura extrudates such as expansion ratio, bulk density, proximate composition and amino acid profile is important to facilitate industrial adoption of this technology. Extruded fura provides consumers with a fast, easy way to prepare which compares favourably with traditional fura. Aesthetic presentation will enhance and increase acceptability of this product. Fura production has potential of increased provision of food especially in the aviation industry, refugee camps, food aids, for areas prone to protein energy malnutrition and those living in war torn famine ravaged areas of west Africa. This product could make a great contribution to food supply in West Africa sub region especially to mitigate the problem of famine in the region.

ACKNOWLEDGEMENT

The funding and support provided by International Foundation for Science (IFS) is gratefully acknowledged and appreciated.

COMPETING INTERESTS

Authors have declared that no competing interests exist.

REFERENCES

Aguilera, J.M., Crisafulli, E.B., Lusas, E.W., Uebersax, M.A., Zabik, M.E. (1984). Air classification and extrusion cooking of navy bean fraction. Journal of Food Science, 49(4), 543 – 548.

Alonso, R., Orue, E., Zabalza, M.J., Grant, G., Marzo, F. (2000a). Effect of extrusion cooking on structure and functional properties of pea and kidney bean proteins. Journal of the Science of Food & Agriculture, 80(3), 397 – 403.

Alonso, R., Aguirre., A., Marzo, F. (2000b). Effect of extrusion and traditional processing methods on antinutrients and in vivo digestibility of protein and starch in faba and kidney beans. Food Chemistry, 68, 159 – 165.

Alvarez – Martinez, L., Kondury, K.P., Harper, J.M. (1988). A general model for 568 expansion of extruded products. Journal of Food Engineering, 53, 609 – 615.

AOAC. (1984). Official method of Analysis, 14th ed. Association of Official Analytical Chemists Washington D.C.

Areas, J.A.G. (1992). Extrusion of food proteins. Crit. Rev. Food Sci. Nutri., 32, 365-392

Box, G.E.P., Hunter, J.S. (1957). Multifactor experimental design for exploring response surfaces. Annals Math. Stat., 28, 195-242.

Bressani, R. (1975). Nutritive value of cowpea. "In Research production and utilization" S.R. Singh , K.O. Rachie (Ed.). John Wiley and Sons, New York.

Chen, J., Serafin, F.L., Pandya, R.N., Daun, H. (1991). Effect of extrusion conditions on sensory properties of corn meal extrudates. Journal of Food Science, 56(1), 84.

Conway, H.F. (1971). Extrusion cooking of cereals and soybeans. I. Food product Development, 5, 27, 29, 31.

Egan, H., Kirk, R.S., Suwyer, R. (1981). Pearson Chemical Analysis of Foods. 2nd Ed. Longman Group London.

FAO/WHO/UNU. (1985). Energy and protein requirements. Tech. Rep. Series 724, Expert Consultation. Geneva: World Health Organization.

FAO/WHO. (1973). Energy and protein requirements. Report of Joint FAO/WHO ad Hoc Expert Committee. WHO Technical Reports series No. 522: Geneva FAO/WHO.

Filli, K.B., Nkama, I. (2007). Hydration properties of extruded fura from millet and legumes. British Food Journal, 109(1), 68-80.

Filli1, K.B., Nkama, I., Abubakar, U.M., Jideani, V.A. (2010). Influence of extrusion variables on some functional properties of extruded millet-soybean for the manufacture of 'Fura': A Nigerian traditional food. African Journal of Food Science, 4(6), 342- 352.

Gujska, E., Czarnecki, Z., Khan, K. (1996). High temperature extrusion of pinto beans (Phaseolus vulgaris) and field pea (Pisum sativum) flours. Polish Journal of Food and Nutrition Science, 46(1), 51 – 60.

Gujral, H.S., Singh, N., Singh, B. (2001). Extrusion behaviour of grits from flint and sweet corn. Food Chemistry, 74, 303 – 308.

Harmann, D.V., Harper, J.M. (1973). Effect of extruder geometry on torque and flow. American Society of Agricultural Engineering, 16, 1175-1178.

Harper, J., Jansen, G. (1985). Production of nutritious precooked foods in developing countries by low – cost extrusion technology. Food Rev. Int., 1, 27 – 97.

Hoseney, R.C., Faubion, J.M., Reddy, V.P. (1992). Organoleptic implications of milled Pearl millet pages 27-32 in utilization of sorghum and millets (Gumez, M.I., House, L.R., Rooney, L.W., Dendy, D.A.V. eds.) Patancheru, A.P. 502 324, India: International Crops Research Institute for the Semi-Arid Tropics.

Ilo, S., Liu, Y., Berghofer, E. (1999). Extrusion cooking of rice flour and amaranth blends. Lebensmittel – Wissenschaft und Technologie, 32, 79 – 88.

Jideani, V.A., Nkama, I., Agbo E. B., Jideani, I.A. (2002). Identification of the hazard and critical control (HACCP) in traditional fura manufacture. Nigerian Food Journal, 19, 42-48.

Launary, B., Lisch, J.M. (1983). Twin-screw extrusion cooking of starches: flow behaviour of starch pastes, expansion and mechanical properties of extrudates. Journal of Food Engineering, 2, 259-280.

Lazou, A., Krokida, M. (2010). Functional properties of corn and corn – lentil extrudates. Food Research International, 43, 609 – 616.

Likimani, T.A., Sofos, J.N., Maga J.A., Harper, J.M. (1991). Extrusion cooking of Corn/soybean mix in presence of thermostable a-amylase. Food Science, 56(1), 99-105.

Nkama, I., Malleshi, N.G. (1998). Production and nutritional quality of traditional Nigerian Masa from mixtures of rice, pearl millet cowpea and groundnut. Food Nutr. Bull., 5(4), 366-373.

Nkama, I., Filli, K. B. (2006). Development and Characteristics of Extruded fura from mixtures of pearl millet and grain legumes flours. International Journal of Food Properties, 9, 157-165.

Mclarem, C.G., Bartolome, V.I., Carrasco, M.C., Quintana, L.C., Ferino, M.I.B., Mojica, J.Z., Olea, A.B., Paunlagui, L.C., Ramos, C.G., Ynalvea, M.A. (1977). Experimental design and data analysis for agricultural research, vol. 1. Los Banos, Laguna: International Rice Research Institute.

Mitzner, K., Shrimshaw, N., Morgan, R. (1984). Improving the nutritional status of children during weaning period. International Food and Nutrition Program. Cambridge, Massachusetts 02139 USA.

Mwasaru, A.M., Muhammad, K., Bakar, J., CheMan, B.Y. (1999). Effects of isolation technique and conditions on the extractability, physicochemical and functional properties of pigeonpea (Cajanus cajan) and cowpea (Vigna unguilata) protein isolates. I. Physicochemical properties. Journal of Food Chemistry, 67, 435 – 443.

Myers, R.H., Montgomery, D.C. (2002). Response Surface Methodology: Process and Product Optimization Using Designed Experiment (2nd ed). John Wiley & Sons. Inc.

Obatolu, V.A. (2002). Nutrient and sensory qualities of extruded malted or unmalted millet/soybean mixture. Food Chemistry, 76, 129-133.

Onyango, C., Henle, T., Zeims, A., Hofmann, T., Bley, T. (2004). Effect of extrusion variables on fermented maize – finger millet blend in the production of uji. LWT, 37, 409 – 415.

Pelembe, L.A.M., Erasmus, C., Taylor, J.R.N. (2002). Development of a Protein – rich Composite Sorghum – Cowpea Instant Porridge by Extrusion Cooking Process. Lebensm – Wiss. U. – Technol., 35, 120 – 127.

Padmanabhan, M., Bhattacharya, M. (1989). Analysis of pressure drop in the extruder dies. J. Food Sci., 54(3), 709 – 713.

Pansawat, Jangchud, K., Jangchud, A., Wuttijumnong, P., Saalia, F.K., Eitenmiller, R.R., Philips, R.D. (2008). Effects of extrusion conditions on secondary extrusion variables and physical properties of fish, rice – based snacks. LWT, 41, 632 – 641.

Qing – Bo, D., Ainsworth, P., Tuker, G., Marson, H. (2005). The effect of extrusion condtions on the physicochemical properties and sensory characteristics of rice – based expanded snacks. Journal of Food Engineering, 66, 284 -289.

Sefa – Dedeh, S., Cornelius, B., Sakyi – Dawson, Afoakwa, E.O. (2003). Application of response surface methodology for studying the quality characteristics of cowpea – fortified nixtamalized maize. Innovative Food Science and Emerging Technologie, 4, 109 -119.

Seker, M. (2005). Selected properties of native starch or modified maize starch/soy protein mixtures extruded at varying screw speed. J. of the Sc. of Food and Agric., 85(7), 1161-1165.

Singh, S., Gamlath, S., Wakeling, L. (2007). Nutritional aspects of food extrusion: a review. International Journal of Food Science and Technology, 42, 916 – 929.

Sopade, P.A., Le Grys, G.A. (1991). Effect of added sucrose on extrusion cooking of maize Starch. Food control, 2, 103-109.

Sotelo, E., Hernandez, M., Montalvo, I., Sausa, V. (1994). Amino acid content and protein biological evaluation of 12 Mexican varieties of rice. Cereal Chemistry, 71, 605.

SPSS. (1998). Statistical package for the social sciences for windows Illinois, USA: SPSS

Sumathi, A., Ushakumari, R., Malleshi, N.G. (2007). Physicochemical characteristics, nutritional quality and shelf – life of pearl millet based extrusion cooked supplementary foods. International Journal of Food Science and Nutrition, 58(5), 350 – 362.

Varnalis, A.I., Brennan, J.G., Macdougall, D.B., Gilmour, S.G. (2004). Optimization of high temperature puffing of potato cubes using response surface methodology. Journal of Food Engineering, 61, 153 – 163.

Vogel, S., Graham, M. (1979). Sorghum and millet: Food Production and use. Report of a Workshop held in Nairobi, Kenya, IDRC, Ottawa, 3 Ontario Canada July, 1978.

Wanasundara, P.K.J.P.D., Shahidi, F. (1996). Optimization of hexametaphosphate-assisted extraction of flaxseed protein using response surface methodology. J. Food Sci., 61(3), 604-607.

Wang, S., Chen, F., Wu, J., Wang, Z., Liao, X., Hu, X. (2007). Optimization of pectin extraction assisted by microwave from apple pomace using response surface methodology. Journal of Food Engineering, 78, 693 – 700.

Wu, M., Li, D., Wang, L., Ozkan, N., Mao, Z. (2010). Rheological properties extruded dispersions of flaxseed – maize blend. Journal of Food Engineering, 98, 480 – 49.

Zasypkin, D.V., Tung- Ching, L. (1998). Extrusion of Soybean and wheat flour as affected by

2

Probiotics and Prebiotics: An Update from the World Gastrointestinal Organization (WGO)

Alfonso Clemente[1*]

[1]Department of Physiology and Biochemistry of Nutrition, Estación Experimental del Zaidín (CSIC), Profesor Albareda 1, 18008 Granada, Spain.

SUMMARY

The human intestine harbors a complex microbial ecosystem consisting of an extraordinary number of resident commensal bacteria existing in homeostasis with the host (Eckburg et al., 2005). This endogenous microbiota establishes a symbiotic mutualistic relationship and impacts on numerous physiological functions including nutrition exchange, control of epithelial cell proliferation/differentiation, pathogen exclusion and stimulation of the immune system (Flint et al., 2007; Cerf-Bensussan et al., 2010). Given the emergent evidence of the roles played by the human microbiota in health and disease, there is a growing interest in identifying live microorganisms (probiotics) and dietary compounds (prebiotics) capable of modulating the composition and metabolic activities of the intestinal microbiota in order to confer beneficial effects on the host. In October 2011, a position paper about probiotics and prebiotics has been published by an Expertise Committee updating the World Gastroenterology Organisation (WGO) Practice guideline
(http://www.worldgastroenterology.org/assets/export/userfiles/Probiotics_FINAL_2011112 8.pdf). In this relevant document, probiotics and prebiotics are defined, being quality, safety and labeling criteria discussed. Information on suppliers of prebiotics and probiotics as well as the main strains used as probiotics in the market are shown. The study also summarizes a number of clinical conditions for which there is evidence for the preventive or therapeutic use of probiotics and prebiotics in pediatric or adult populations.
Probiotics are defined as live microorganisms that confer a health benefit on the host when administrated in adequate amounts (FAO/WHO 2002). The probiotics are formulated in foods, drugs and dietary supplements. Species of the genera *Bifidobacterium* and *Lactobacillus* are amongst the species most commonly used; both are believed to play an important role in maintaining and promoting a healthy gut environment. The main health benefits of probiotics are associated with regulation of the intestinal tract microbiota by

*Corresponding author: Email: alfonso.clemente@eez.csic.es

reducing the numbers or colonization of pathogenic bacteria, maintenance of the epithelial cell integrity and barrier function as well as induction of immunoregulatory mechanisms that control adaptive immune functions (van Loveren et al., 2012). As reported in the WGO document, the health benefits and efficacy of probiotics can be only attributed to specific strains, timing of administration and dosage previously tested in human trials; however, the number of studies assessing these variables in a comparative manner is rather limited. The vehicle/filler in delivering probiotics plays a critical role given that can significantly affect the microbial viability rates. Other relevant aspects like storage and safety under the conditions of recommended use should be also considered. In last years, a number of potential health benefits associated to probiotics have been reported. In order to support such claims, well-designed double-blind, placebo-controlled human trials are strictly necessary. The strongest clinical evidence for probiotics is related to their use in improving gut health and stimulating gut function, being well documented their beneficial effects on a variety of intestinal disorders. Probiotics affect the intestinal microbioma by stimulating immune and non-immune mechanisms through antagonism and competition with potential pathogens. Several probiotics strains have been reported to be effective in reducing the severity and duration of infectious diarrhea in children (Szajewska et al., 2007a, 2007b), being some specific strains effective in the prevention of antibiotic-associated diarrhea (Vanderhoof et al., 1999; Correa et al., 2005; Ruszczynski et al., 2008). Some probiotic strains have demonstrated to be effective in the prevention of pouchitis (Gionchetti et al., 2003) and remission of ulcerative colitis (Kruis et al., 2004), but not for maintenance of remission in Crohn´s disease. Regarding irritable bowel syndrome (IBS), there are some clinical evidences about the alleviation of abdominal pain after probiotic treatment (Moayyedi et al., 2010). In addition, *Streptococcus thermophilus* and *Lactobacillus delbrueckii* subsp. *bulgaricus* seems to improve lactose digestion and reduce symptoms related to lactose maldigestion (EFSA, 2010). Although probiotics reduces the risk of colon cancer in animal models, their chemopreventive effects in humans have not been proven yet.

Prebiotics have been recently redefined as 'non-digestible functional ingredients which are selectively fermented and allow specific changes, both in the composition and/or activity of the gastrointestinal microflora that confers benefits upon host well-being and health' (Roberfroid, 2007). They favor the growth of beneficial bacteria over that of harmful ones. Among intestinal bacteria stimulated by prebiotics, *Bifidobacterium spp.* and *Lactobacillus spp.* are amongst the species most relevant. The major prebiotic oligosaccharides on the market are fructan inulin, lactulose, fructo-oligosaccharides (FOS) and galacto-oligosaccharides (GOS) (Rastall, 2010). It is generally accepted that the major beneficial effects of prebiotic carbohydrates occur in the large intestine due to the slow transit of the substrates to be fermented and their effects on microbiota diversity which plays an important role in host health (Gibson, 2004). Fermentation of some of these prebiotics in large intestine is reported to exert a positive effect on growth rates of several strains of *Bifidobacterium* and/or *Lactobacillus* (Cardelle-Cobas et al., 2009; 2011); other potential mechanisms might include receptor blockade or production of health-promoting components.

Health claims are conceived to support healthy European consumers to make healthy food choices in order to develop healthy lifestyles (van Loveren et al., 2012). Despite the substantial amount of basic and clinical research on the beneficial effects of probiotics and prebiotics, to date none of the claims submitted to the European Food Safety Authority (EFSA) and reviewed by the Panel on Dietetic Products, Nutrition and Allergies (NDA) have been accepted (Guarner et al., 2011). There are several reasons for health claims to be rejected, such as insufficient characterization of the probiotic strain/s or prebiotic carbohydrate, lack of suitable human intervention studies to validate the claim for the intended population group, or a need for identification of novel biomarkers to assess cause and effect between consumption of the probiotic/prebiotic and the claimed health effects, among others (Aggett et al., 2005; Verhagen et al., 2010). However, health claims supported by robust and solid evidence are also being rejected. Some concern in the scientific community about inconsistencies in the nature of the opinions of EFSA as well as its lack of clarity on the criteria to substantiate health claims, from study design to wording, have been recently reported (O´Connor, 2011; Guarner et al., 2011). In addition, there is growing disparity in scientific opinion over whether changes in the number of nonpathogenic microorganisms can be viewed as a beneficial marker for digestive and immune health. Regarding this, EFSA have published a guidance document on scientific requirements for health claims related to gut and immune function to facilitate study design for submissions (EFSA, 2011). EFSA claims that changes in gastrointestinal microbiota should be accompanied by a beneficial physiological or clinical outcome. The use of molecular analytical platforms as well as the identification of biomarkers and/or parameters related to gut and immune function would help us to determine novel criteria for developing effective probiotics and prebiotics. This effort will increase the protection of consumers from misleading or untruthful health claims but also should stimulate innovation in the food industry to offer a wider range of healthier foods to consumers (Buttriss, 2010; van Loveren et al., 2012).

Keywords: Gastrointestinal health; microbiota; prebiotics; probiotics.

ACKNOWLEDGMENTS

The author acknowledge support from ERDF-co-financed grant from Junta de Comunidades de Castilla-La Mancha (POII10-0178-4685).

COMPETING INTERESTS

Author has declared that no competing interests exist.

REFERENCES

Aggett, P.J., Antoine, J.M., Asp, N.G., et al. (2005). PASSCLAIM: consensus on criteria. European Journal of Nutr., 44, i5-30.

Buttriss, J.L. (2010). Are health claims and functional foods a route to improving the nation´s health?. Nutrition Bull., 35, 87-91.

Cardelle-Cobas, A., Fernández, M., Salazar, N., Martinez-Villaluenga, C., Villamiel, M., Ruas-Madiedo, P., de los Reyes-Gavilan, C. (2009). Bifidogenic effect and stimulation of short chain fatty acid production in human faecal slurry cultures by oligosaccharides derived from lactose and lactulose. Journal of Dairy Res., 76, 317-25.

Cardelle-Cobas, A., Corzo, N., Olano, A., Pelaez, C., Requena, T., Avila, M. (2011). Galactooligosaccharides derived from lactose and lactulose: influence of structure on *Lactobacillus*, *Streptococcus* and *Bifidobacterium* growth. International Journal of Food Microbiol, 149, 81-87.

Cerf-Bensussan, N., Gaboriau-Routhiau, V. (2010). The immune system and the gut microbiota: friends or foes? Nature Reviews Immunol, 10, 735-44.

Correa, N.B., PeretFilho, L.A., Penna, F.J., Lima, F.M., Nicoli, J.R. (2005). A randomized formula controlled trial of *Bifidobacterium lactis* and *Streptococcus thermophiles* for prevention of antibiotic-associated diarrhea in infants. Journal of Clinical Gastroenterol, 39, 385-389.

Eckburg, P.B., Bik, E.M., Bernstein, C.N., Purdom, E., Dethlefsen, L., Sargent, M., Gill, S.R., Nelson, K.E., Relman, D.A. (2005). Diversity of the human intestinal microbial flora Science, 308, 1635-1638.

EFSA Panel on Dietetic Products, Nutrition and Allergies (NDA). (2010). Scientific opinion on the substantiation of health claims related to live yoghurt cultures and improved lactose digestion (ID 1143, 2976) pursuant to Article 13(1) of Regulation (EC No 1924/2006. EFSA J., 8, 1763.

EFSA Panel on Dietetic Products, Nutrition and Allergies (NDA). (2011). Guidance on the scientific requirements for health claims related to gut and immune function. EFSA J., 9, 1984.

FAO/WHO. (2002) Guidelines for the evaluation of probiotics in food. Report of a joint FAO/WHO Working Group.ftp://ftp.fao.org/docrep/fao/009/a0512e/a0512e00.pdf.

Flint, H.J., Duncan, S.H., Scott, K.P., Louis, P. (2007). Interactions and competition within the microbial community of the human colon: links between diet and health. Environmental Microbiol, 9, 1101-1111.

Gibson, G.R. (2004). From probiotics to prebiotics and a healthy digestive system. Journal of Food Sci., 69, 141-143.

Gionchetti, P., Rizzello, F., Helwig, U. et al. (2003). Prophylaxis of pouchitis onset with probiotic therapy: a double-blind, placebo-controlled trial. Gastroenterology, 124, 1202-1209.

Guarner, F., Sanders, M.E., Gibson, G. et al. (2011). Probiotic and prebiotic claims in Europe: seeking a clear roadmap. The British Journal of Nutr., 106, 1765-1767.

Kruis, W., Fric, P., Pokrotnieks, J., et al. (2004). Maintaining remission of ulcerative colitis with the probiotic *Escherichia coli*Nissle 1917 is as effective as with standard mesalazine. Gut., 53, 1617-1623.

Moayyedi, P., Ford, A.C., Talley, N.J., et al. (2010). The efficacy of probiotics in the treatment of irritable bowel syndrome: a systematic review. Gut., 59, 325-332.

O'Connor, A. (2011). Nutrition and health claims- where are we with the health claims assessment process? Nutrition Bull., 36, 242-247.

Rastall, R.A. (2010). Functional oligosaccharides: application and manufacture. Annual Reviews of Food Science and Technol., 1, 305-339.

Roberfroid, M. (2007). Prebiotics: the concept revisited. Journal of Nutr., 137, 830-837.

Ruszczynski, M., Radzikowski, A., Szajewska, H. (2008). Clinical trial: effectiveness of *Lactobacillus rhamnosus* (strains E/N, Oxy and Pen) in the prevention of antibiotic-associated diarrhea in children. Alimentary Pharmacology & Ther., 28, 154-161.

Szajewska, H., Ruszczynski, M., Gieruszczak-Bialek, D. (2007). *Lactobacillus* GG for treating acute diarrhea in children.A meta-analysis of randomized controlled trials. Alimentary Pharmacology & Ther., 25, 177-184.

Szajewska, H., Skorka, A., Dylag, M. (2007). Meta-analysis: *Saccharomyces boulardii* for treating acute diarrhea in children. Alimentary Pharmacology & Ther., 25, 257-264.

Vanderhoof, J.A., Whitney, D.B., Antonson, D.L., Hanner, T.L., Lupo, J.V., Young, R.J. (1999). *Lactobacillus* GG in the prevention of antibiotics-associated diarrhea in children. Journal of Pediatr., 135, 564-568.

van Loveren, H., Sanz, Y., Salminen, S. (2012). Health claims in Europe: probiotics and prebiotics as case examples. Annual Review of Food Science and Technol. (doi:10.1146/annurev-food-022811-101206).

Verhagen, H., Vos, E., Francl, S., Heinonen, M., et al. (2010). Status of nutrition and health claims in Europe. Archives of Biochemistry et Biophys., 501, 6-15.

3

Risk-Benefit Assessment of Cold-Smoked Salmon: Microbial Risk versus Nutritional Benefit

Firew Lemma Berjia[1*], Rikke Andersen[2], Jeljer Hoekstra[3], Morten Poulsen[4] and Maarten Nauta[1]

[1]Division of Epidemiology and Microbial Genomics, The National Food Institute, Technical University of Denmark, Mørkhoj Bygade 19, 2860 Søborg, Denmark.
[2]Division of Nutrition, The National Food Institute, Technical University of Denmark, Mørkhoj Bygade 19, 2860 Søborg, Denmark.
[3]National Institute of for Public Health and the Environment (RIVM) Bilthoven, The Netherlands.
[4]Division of Toxicology and Risk Assessment, The National Food Institute, Technical University of Denmark, Mørkhoj Bygade 19, 2860 Søborg, Denmark.

Authors' contributions

This work was carried out in collaboration between all authors. Author FLB designed the study, did literature research, developed the model, performed the statistical analysis, and wrote the first draft of the manuscript. The other authors contributed by discussing the model, adding expertise from various research disciplines and helping in finalizing the paper. In addition, authors MN and JH assisted in the model development and statistical analysis. All authors read and approved the final manuscript.

ABSTRACT

The objective of the study is to perform an integrated analysis of microbiological risks and nutritional benefits in a fish product, Cold Smoked Salmon (CSS).
Literature study identified the major health risks and benefits in connection with CSS consumption. The reduction of the risk of Coronary Heart Disease (CHD) mortality and stroke, as well as enhanced cognitive (IQ) development of unborns following maternal intake, are identified as the main health benefits of omega-3 fatty acid from CSS. Contrary, risk of meningitis, septicemia and abortion/stillborn are identified as a major health risk endpoints due to exposure to the pathogen *L. monocytogenes*.

Corresponding author: Email: fber@food.dtu.dk

Two consumption scenarios were considered: a reference scenario (23g/day and 20g/day for man and woman respectively) and an alternative scenario (40g/day for both sexes). In order to evaluate and compare the risks and benefits, the Disability Adjusted Life Years (DALY) method has been used as a common metric.

Results show that the overall health benefits outweigh the risk, foremost contributed by the effect of decreased CHD mortality and IQ increase. A sensitivity analysis indicated that this result was robust for the analyzed parameters, except the storage time: the adverse effect of consumption of CSS prevails over the beneficial effect if the storage time of CSS is increased from two weeks to five weeks or more, due to an increased risk of listeriosis. This study demonstrates how microbial risks can be integrated in risk-benefit assessment, and shows that a sensitivity analysis has an added value, even if the benefits largely outweigh the risk in the initial analysis.

Keywords: Cold-smoked salmon; Listeria monocytogenes; omega-3 fatty acids; DALY.

1. INTRODUCTION

Risk-benefit assessment is the weighing of the probability of an adverse health effect against the probability of a beneficial effect as a result of exposure/intake of food (EFSA, 2010). Examples and a guidance of how to perform risk-benefit assessment of foods have recently been provided (Hoekstra et al., 2010; EFSA, 2010). Nonetheless, risk-benefit methods need further development. There is currently no internationally agreed method to perform human health risk-benefit assessment of food and so far only a few risk-benefit assessments studies included microbiological hazards (Havelaar et al., 2000; Magnússon et al., 2012). Typical aspects of microbiological risk assessment, like the inclusion of the impact of storage and processing on the weighing of the risk and benefit, are therefore rarely included in published risk-benefit assessments.

In this paper we present a risk-benefit assessment on a fish product. Several studies have assessed the risk of toxic contaminants and benefits of nutrients following the consumption of fish (Gladyshev et al., 2008; Cohen et al., 2005b; Guevel et al., 2008; FAO/WHO, 2011; Hoekstra et al., 2012) and found that in general the public health benefits are larger than the risks. However, microbial risks have not been integrated into these risk-benefit assessments. The present study aims to illustrate how a microbiological hazard can be included in a typical risk-benefit assessment and how this may add to the existing risk-benefit assessment tools and methodologies. Furthermore, we included a sensitivity analysis to evaluate the impact of some of the model parameters on the assessment.

2. RISK-BENEFIT ASSESSMENT OF COLD-SMOKED SALMON: MODEL

2.1 Scope

The risk of the bacterial pathogen (*L. monocytogenes*) is evaluated against the benefits of the intake of omega-3 fatty acids in a risk-benefit assessment of CSS consumed in Denmark. Salmon is an oily fish containing considerable amount of omega-3 fatty acids, it is a popular ready-to-eat food in most part of the world and it is consumed in many European countries (WHO/FAO, 2004).

The assessment compares a reference scenario with an alternative scenario, as in (Hoekstra et al., 2010). In this comparison it is assumed that CSS is added to the normal diet in an isocaloric way and substitution by other food items with potential health effects is neglected. Best estimates are applied for the various model parameters. Disability Adjusted Life Years (DALYs) are used as an integrated health measure to compare risks and benefits (Hoekstra et al., 2010).

All statistical and mathematical modelling is implemented in Microsoft Office, version 2007 except for the dose-response modelling of CHD mortality and stroke (Appendix) which was performed on statistical software R, version 2.10.1.

2.2 Hazard Identification, Selected Health Effects and Affected Subpopulation

Cold-smoked fish products may be contaminated with *L. monocytogenes,* the agent that causes foodborne listeriosis. Vacuum-packed cold-smoked fish has a long shelf-life and can support the growth of *L. monocytogenes* (WHO/FAO, 2004). The contribution of salmon for the cases of listeriosis has been reported (Pouillot et al., 2007; Lindqvist and Westoo, 2000; WHO/FAO, 2004). Recently an increasing incidence of invasive listeriosis, primarily septicemia and meningitis have been reported in several European countries (Allerberger and Wagner, 2010; Jensen et al., 2010). Listeriosis during pregnancy is also a serious threat to the unborn child, which can lead to abortion/ stillborn (Smith, et al., 2009). Hence, meningitis, septicemia and abortion/stillborn are selected as the endpoints following exposure to *L. monocytogenes*.

Elderly, immunocompromized and pregnant women and/or their unborn fetuses are the most susceptible groups for listeriosis (WHO/FAO, 2004; Allerberger and Wagner, 2010). Therefore, both sexes aged ≥ 60 are selected for septicemia and meningitis. The population of interest for abortion/stillborn, are potentially pregnant women aged 20-45.

2.3 Benefit Identification, Selected Health Effects and Affected Subpopulation

The nutrients in fish that have plausible and significant health benefits for human are omega-3 fatty acids, principally eicosapentaenoic acid (EPA) and docosahexanoic acid (DHA) (Mozaffarian, 2006). Intake of fish may protect against CHD mortality and stroke (Mozaffarian, 2006). In addition, an association is found between the maternal intake of DHA and a beneficial effect in cognitive development of their unborn child, measured as an increase in IQ (Cohen et al., 2005a). Zeilmaker et al. (2012) investigated both the adverse (MeHg) and beneficial (DHA) effect of fish intake on IQ and found a very small IQ gain for salmon intake. Therefore, reduction of CHD mortality and total stroke, and improved cognitive development are selected endpoints in this paper.

Most of the studies mentioned in the Appendix (Table 8 and 9), which are incorporated in the dose-response modeling of CHD mortality and total stroke, included adults of both sexes older than 35 years. Hence, both sexes aged ≥ 35 are selected for both endpoints as a target population. For the benefit of maternal intake of DHA on the child's IQ, it is assumed that women aged 20-45 give birth with different probabilities depending on age (Table 3).

2.4 Intake and Exposure Assessment

The current mean fish intake is 20 and 23 g of fish/day for women and men respectively (Pedersen et al., 2010). In the reference scenario it is assumed that every individual consumes the current mean fish consumption as CSS i.e. 20 and 23 g for women and men. For the alternative scenario it is assumed that every person consumes 40 g CSS/day. The intake of CSS is assumed the same for all age groups in each scenario.

Omega-3 fatty acid intake is computed from the official Danish Food Composition Database in combination with the consumption scenarios (Denmark Technical University, 2011).

For *L.monocytogenes*, an exponential growth model is applied to assess the distribution of concentrations at consumption as a function of initial concentration, storage time, growth rate and lag-time (Table 1, eq. 7). The 10-based logarithm of initial concentrations (N_0) of *L.monocytogenes* {0.5: 1.5: 2.5: 3.5} and their prevalences {0.28: 0.05: 0.01: 0} are taken from Jørgensen and Huss (1998). For the exposure assessment, these results are combined with the consumption scenarios.

2.5 Integration of the Health Effects

To combine the health outcomes of the risk and the benefit we have chosen the DALY model of (Hoekstra et al., 2010). For an individual of age CA, the amount of DALY per person per year is:

$$DALY_{a,s}=P_{eff,a,s}[(P_{rec}*YLD_{rec}*w+P_{die}(YLD_{die}*w+LE_{a,s}-CA-YLD_{die})+(1-P_{die}-P_{rec})*(LE_{a,s}-CA)*w]$$

Where:
$DALY_{a,s}$	disability adjusted life years at age, *a* and sex, *s*
$P_{eff,a,s}$	probability of onset of the disease at age, *a* and sex, *s*, per year
P_{rec}	probability of recovery from the disease
P_{die}	probability the disease causes death
YLD_{rec}	duration of disease for those who recover
YLD_{die}	duration of disease for those who die
CA	current age of individual in year of disease onset (years)
$LE_{a,s}$	normal life expectancy for an individual of age CA[1]
w	disability weight for disease.

2.6 Dose-Response Relationship

After a literature survey, eleven and eight studies are incorporated for the dose-response modeling of CHD mortality and total stroke respectively (Appendix, Tables 8 and 9). The results from studies that are included in the dose-response relation of CHD mortality and total stroke have been implemented by a relation where the relative risk (RR) of the health outcomes is a function of fish intake. Different functions are analyzed in order to select the best model based on the best fit statistics (Appendix 1).

---

[1] LE is interpreted here as expected age at death not as the also commonly used expected remaining years of life

Table 1. Model equations applied and point estimates of the parameters

Model equation	Parameter values	Description (unit)
1. $I_{DHA} = F_{intake} * DHA$	F_{intake}, scenarios $DHA = 1.16g/100g$	I_{DHA}, intake of DHA (g/d); $F_{intake,}$ fish intake (g/d); DHA content of CSS (g DHA per g CSS) (DTU, 2011).
2. $\Delta IQ = d*I_{DHA}$ (Cohen et al. 2005a)	d = 1.3, uncertainty interval (0.8-1.8)	ΔIQ, change in intelligent quotient; d, coefficient.
3. $P_{eff(IQ)}$ = Probability of a woman giving a birth,	$P_{eff(IQ)}$, vary depending on age of a women giving a birth	$P_{eff(IQ)}$, probability of onset of IQ effect which is equivalent to probability of a woman giving a birth (Table 3).
4. $\ln(RR) = a + b* \ln(F_{intake})$	RR, a=0.17, b=0.137 for CHD mortality and a=0.113, b=0.094 for stroke	RR, relative risk of CHD mortality and total stroke; a and b are estimated in the meta-analysis in appendix1.
5. $P_{eff,ep,r\,(a,s)} = Inc_{(a,s)}/N_{(a,s)}$	Variable with age a, sex s and endpoint ep	$P_{eff,ep,r(a,s)}$ probability of onset of endpoints ep, (CHD mortality and total stroke) at reference intake r, in 5 year age class a, for sex s; $Inc_{(a,s)}$, current incidence of endpoint ep in (a,s); $N_{(a,s)}$, number of population in (a,s). Note: It is assumed that the probability of effect is the current incidence rate for reference intake for both endpoints.
6. $P_{eff,ep,a(a,s)} = RR_{(s)a} \times P_{eff,\,ep,r\,(a,s)}/RR_{(s)r}$	Varies with scenarios, ages and sexes	$P_{eff,ep,a(a,s)}$, probability of onset of endpoint at alternative intake at age and sex; $RR_{(s)a}$, relative risk of alternative scenario at sex, s; $RR_{(s)r}$, relative risk of reference scenario at sex, s.
7. $\log N_t = \log N_0 + \mu(t - \lambda)$	N_0, (see section 2.4), t=14; μ=0.113; λ=0.167 (WHO/FAO, 2004)	N_t, concentration of *Listeria* after storage (CFU/g); N_0, initial concentration (CFU/g); μ, growth rate (log CFU/d); t, storage time (day); λ, lag-time (day). At storage temperature of 5°C
8. $D_{listeria} = F_{intake} * N_t$		$D_{listeria}$, dose of listeria (CFU/d).
9. $P_{inf} = 1-e^{-rD_{listeria}}$	r= 5.85 * 10^{-12}	P_{inf}, probability of infection and illness of *Listeria; r*, dose-response parameter specific to *Listeria* for susceptible population (WHO/FAO, 2004).
10. $P_{eff(mengi)} = K_{mengi} * P_{inf}$	K = 0.24	$P_{eff(mengi)}$, probability of onset of meningitis; K_{mengi}, proportion of meningitis cases among those infected with *Listeria*.
11. $P_{eff(septi)} = K_{septi} * P_{inf}$	K_{septi} = 0.74	$P_{eff(septi)}$, probability of onset of septicemia; K_{septi}, proportion of septicemia cases among those infected with *Listeria*.
12. $P_{eff(abo/stl)} = K_{(abo/stl)} * P_c * P_{inf}$	$K_{(abo/stl)}$ = 0.266; P_c, variable with age	$P_{eff(abo/stl)}$, probability of onset of abortion/stillborn; $K_{abo/stl}$, proportion of abortion/stillborn among pregnant women infected with *Listeria*. P_c, probability of giving birth

In addition, the models are validated using residual analysis and QQ-plot (Ekstrøm and Sørensen, 2011). Based on this, the log-linear model has been selected to estimate the RR based on the intake scenarios. Then, the estimated RR are converted into absolute risk by combining the RR's and the current incidence rates of CHD mortality and total stroke, which are obtained from Denmark Statistics (2011). To characterize the benefit of maternal DHA intake to the cognitive development (IQ) of their offspring, the relation described in Table 1, eq. 1 is applied.

The exponential dose-response model for *L. monocytogenes* is used to characterize and estimate the probability of infection (Table 1, eq. 9).

The DALY calculation has been performed for each sex, age and scenario. From these, the total DALYs are calculated for the Danish population by summation; population's data are shown in Table 2 and obtained from Denmark Statistics (2011). Parameter estimates for DALY computation are also presented in Table 5 and explained in sections 2.7 and 2.8.

Table 2. Number of population by age and gender (Denmark Statistics, 2011)

		Sex		
		Man	Woman	Total
Age	≥ 60	582589	642706	1225295
	18-49		1155573	1155573
	≥ 35	1556888	1631905	3188793
				5569661

2.7 Estimation of DALY Parameters for Listeriosis

The probability of developing septicemia $(P_{eff(septi)})$ and meningitis $(P_{eff\,(mengi)})$ depends on the infection probability, P_{inf}, which depends on fish intake. In the reference scenario P_{inf} is $8.9*10^{-6}$ for women (20g CSS) and $1*10^{-05}$ for men (23 g CSS. In the alternative scenario (40 g CSS) it is $1.78*10^{-05}$. In this study it is assumed that the percentage of septicemia and meningitis (K_{septi}, K_{mengi}) is 74% and 24% respectively (Gerner-Smidt et al., 2005). Studies reported that the percentage of abortion/stillborn ($K_{abo/still}$) is about 15-25% (Mylonakis et al., 2002) and 33.3% (Smith et al., 2009). Consequently, we take the mean (26.6%) of the two reported percentages to estimate the abortion/stillborn percentage.

P_{die} is assumed to equal published case fatality rates, 20.8% and 25.4% of the patients died of septicemia and meningitis respectively within a month of diagnosis in a 10-years follow up study period (Gerner-Smidt et al., 2005). For septicemia it is assumed that people who do not die will recover, so P_{rec} = 1- P_{die}. For meningitis the sequela is taken into account, so P_{rec} = 1- P_{die} -0.14 (Aouaj et al., 2002).

YLD_{die} is computed for meningitis and septicemia from (Gerner-Smidt, et al., 2005). YLD_{rec} and w for both meningitis and septicemia are obtained from (Kemmeren et al., 2006).

Abortion/stillborn implies that the life of a newborn is lost. Therefore, YLD_{die}, YLD_{rec}, CA and P_{rec} are 0 and P_{die} is 1.Obviously abortion/stillborn can only happen with pregnant women therefore the probability of a pregnancy, P_c is included and $P_{eff(abo/stl)}$ is estimated using eq. 12 on Table 1. It is assumed that women give birth with different probabilities depending on age. The probabilities of pregnancy at age below 20 and above 45 are assumed zero.

Table 3. The annual probability that a woman gives birth depending on age (Denmark Statistics, 2011)

Age mother	P_c, Probability of giving birth
20 - 25 year	0.039
25 - 30 year	0.1139
30 - 35 year	0.127
35 - 40 year	0.057
40 - 45 year	0.01

2.8 Estimation of DALY Parameters for CHD Mortality, Stroke and IQ

The case fatality rate of total stroke is assumed to be the same as for ischemic stroke, so P_{die}=0.26 (Andersen et al., 2009). P_{rec} and YLD_{rec} are set to zero, assuming no one can recover from stroke. We also assumed that the YLD_{die} is associated with highest mortality period which is within 30 days, this leads to an estimate of approximately 0.082 (Ingall, 2004; Andersen et al., 2009). The disability weight of stroke varies depending on the stages of stroke. WHO estimated that for the first-ever stroke cases and long-term stroke survivors, w is 0.92 and 0.266 respectively (WHO, 2008). In our case, we take the rounded mean of the two values (w=0.6). P_{eff} depends on RR, age and sex (eq. 4 in Table 1). RR for stroke for 20, 23 and 40 g fish/day is 0.84, 0.83 and 0.79.

For fatal CHD, because no one recovers from fatal CHD, P_{rec} and YLD_{rec} are set to 0. By definition P_{die} of fatal CHD is 1. P_{eff} depends on RR and age and sex (eq. 4 in Table 1). RR for fatal CHD for 20, 23 and 40 g fish/day is 0.79, 0.77 and 0.71 respectively.

For IQ, $P_{eff(IQ)}$ is assumed to be the probability that a woman delivers a baby (P_c, Table 3). The probability of having a particular IQ (from the definition of IQ, normal distributed, mean 100, standard deviation 15) resulting from the change in IQ obtained from Table 1 eq. 2 linked with the disability weight of a particular IQ (Table 4) results in a weighted average w depending on IQ change as in (Hoekstra et al., 2012). For the IQ effect, the parameters YLD_{die}, YLD_{rec}, P_{die}, P_{rec} and CA are 0.

Table 4. Disability weights of IQ levels (Stouthard et al., 1997)

IQ	W
>85	0
70-84	0.09
50-69	0.29
35-49	0.43
20-34	0.82
0 <20	0.76

Table 5. Parameter values for the DALY calculations as estimated from epidemiological data

Health effects	Estimated parameters				
	P_{rec}	YLD_{rec}	P_{die}	YLD_{die}	W
Meningitis	0.625	0.5	0.254	0.08	0.32
Septicemia	0.792	0.02	0.208	0.08	0.93
Abortion/stillborn	0.0	-	1	-	-
CHD mortality	0.0	-	1	-	-
Total stroke	0.0	-	0.26	0.082	0.6
IQ	0.0	-	0.0	-	X^*

X^* is dependent on IQ (Table 4) which depends on the maternal intake of DHA (Table 1 eq 2).

The net DALY is calculated using:

$$\Delta DALY = \sum DALY_{alt} - \sum DALY_{ref}$$

Where, $\Delta DALY$ is change in DALY; $\sum DALY_{alt}$, summation over all persons in the population of DALY's for the alternative scenario; $\sum DALY_{ref}$, summation over all persons in the population of DALY's for the reference scenario.

DALY represents health loss; therefore, if the estimation of $\Delta DALY$ results in a positive value then the change in consumption has an adverse health effect. If the $\Delta DALY$ is negative, then the change in consumption has a beneficial effect (Hoekstra et al., 2010).

2.9 Sensitivity Analysis

A sensitivity analysis is performed to explore the impact of modifying some of the model parameter estimates on the risk-benefit assessment. Targeted parameters are the d-value for the effect of DHA intake on IQ change (Table 1, eq. 2), the parameters a and b defining RR of CHD and total stroke (Table 1, eq. 4), the storage time t (Table 1, eq. 7) and the lag-time of *L. monocytogenes*, λ (Table 1, eq. 7). These parameters relate to different endpoint and are known to be variable and/or uncertain. Moreover, for the estimation of RR of CHD mortality and total stroke one more function (exponential function, $\ln(RR) = b^* F_{intake}$ is analysed to see the difference in DALY estimate of the two endpoints compared to the DALY estimate obtained from function, $\ln(RR) = a + b^*\ln(F_{intake})$.

For example, the storage time of CSS was wide in range in various studies (Hansen et al., 1998; Leroi et al., 2001; WHO/FAO, 2004). *L.monocytogenes* relative lag time in foods is in the range of 0–40h, with a peak value near 2.5. Lag-times in laboratory broths had a similar range, but the peak value was nearer to 4.5h (Ross, 1999). In this study 4h is selected as a baseline lag-time value and converted to day unit (Table 1, eq. 7) and for the sensitivity analysis 2.5h and 4.5h is used from the peak value of foods and laboratory broths.

Furthermore, the uncertainty interval of the d-value in Table 1, eq, 2 and the 95% CI of the selected (bold) function in the appendix on Table 10 for the parameter a and b of RR of both CHD mortality and stroke are analyzed for sensitivity. The sensitivity analysis is done by varying one variable at the time (OAT) while keeping the others constant at their baseline value. More sophisticated sensitivity methods are possible (Saltelli et al., 2000), but in this relatively simple model the OAT approach is sufficient to identify the greatest sources of uncertainty and their approximate influence on the end result.

3. RISK-BENEFIT ASSESSMENT OF COLD-SMOKED SALMON: RESULTS

3.1 Baseline

The assessment shows that increasing the consumption of CSS has an overall health gain with respect to the selected endpoints, as the beneficial effects of fatty acids clearly outweigh the adverse health effect of *Listeria* (Table 7).

The extra cases of the hazardous endpoint and the prevented cases of the beneficial endpoint due to the change in consumption are presented in Table 6 below.

Table 6. The number of extra/prevented cases when change in consumption per year

Endpoints	Reference	Alternative	Extra/prevented cases
Septicemia	8.66	16.2	7.54
Meningitis	2.83	5.25	2.42
Abortion/stillborn	1.5	3	1.5
CHD mortality	5435	4953	-482
Stroke	3787	3580	-207

The number shows the number of cases per year at the different scenario. The last column shows that the additional cases (positive value) due to listeriosis and the prevented number of cases (negative value) due to omega-3 fatty acid when change in consumption.

When comparing the hazardous endpoints, for listeriosis there are more life years lost due to septicemia and meningitis in women compared to men. This is due to a larger increase in intake of CSS for women compared to men. The amount of healthy life years lost is largest for septicemia, followed by abortion/stillborn and meningitis.

Table 7. The baseline DALY's for each sex and scenario

	Men			Women			
	Ref	Alt	ΔDALY	Ref	Alt	ΔDALY	Sum of DALY
Septicemia	13	23	**9.7**	14.6	29	**14.5**	24.
Meningitis	6	10.6	**4.5**	6.74	13.5	**6.8**	11
Abortion/Stillborn*				11	23	**12**	12
CHD mortality	32093	29592	**-2501**	23402	21032	**-2370**	**-4871**
Stroke	11874.7	11302	**-572**	15192.5	14284	**-908**	**-1430.5**
IQ*				-3139	-6181	**-3042**	**-3042**
Net DALY							**-9343**

Ref, reference scenario; Alt, alternative scenario
*Because abortion/stillborn and IQ endpoints are result of maternal consumption on their fetus, the DALY is only reported for women.

On the other hand, there is a large gain in healthy life years for both sexes due to reduction of CHD mortality and stroke. Likewise, a large benefit is obtained due to the IQ effect.

Women achieve more benefit than men by the prevention of stroke and men attain more benefit from the prevention of CHD mortality than women.

As a result, the net public health effect of the change of consumption of CSS leads to a gain of 9343 healthy life years in the population of approximately 5570000.

3.2 Sensitivity Analysis

The sensitivity analysis shows quantitative and qualitative changes in ΔDALY depending on theparameters. As illustrated in Fig. 1, increasing the storage period leads to higher risks of listeriosis and has no effect on fatal CHD, stroke and IQ change in newborns. The net DALY shows that the adverse effect of consumption of CSS prevails over the beneficial effect from five weeks of storage time on. It leads to a loss of 1677 and 58391 healthy life years in the population at five and six weeks storage period respectively.

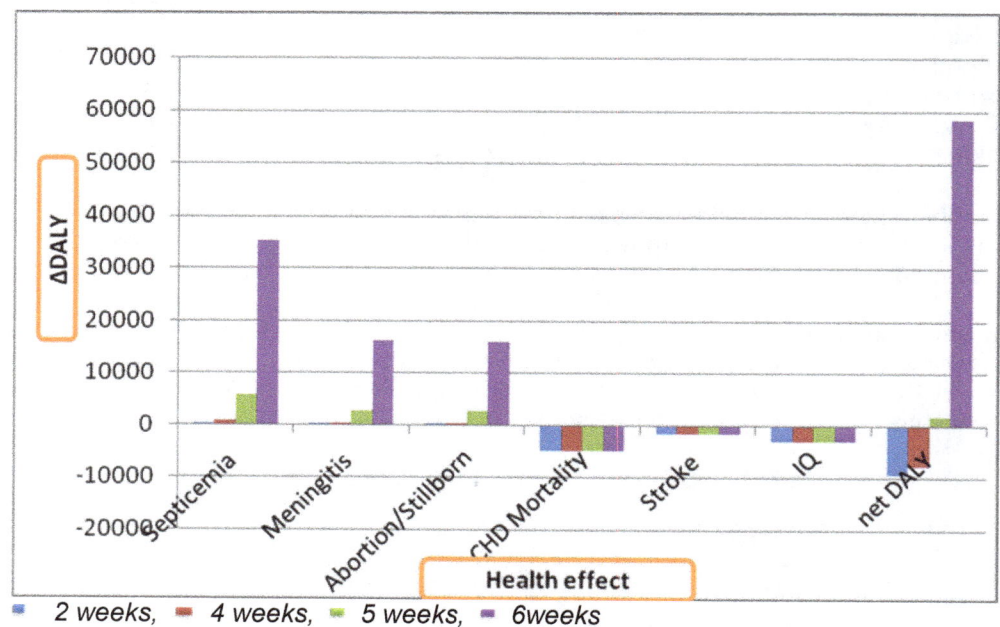

Fig. 1. Sensitivity analysis for the effect of storage time.

The sensitivity analysis for the RR for CHD mortality and total stroke, lag-time of listeria and the d-value uncertainty interval shows no change in the overall balance of the risk-benefit. The result of all the parameters analysed for sensitivity is presented in Fig. 2 below, including the net DALY when using the exponential RR model of CHD mortality and total stroke as an alternative model. The figure includes the net DALY for the baseline scenario given in Table 7, which is represented by "baseline". The various parameters and their values that are used to estimate the baseline net DALY are presented in Table 1.

In Fig. 2 it appears that there is no difference between the resulst of the lag-time sensitivity analysis and the baseline result. The other parameters (RR and d) show a beneficial effect of net DALY which is similar to baseline result qualitatively. However, there can be seen a shift of beneficial to hazardous effect at which the bar extends to the positive direction at storage time of 5 weeks and further.

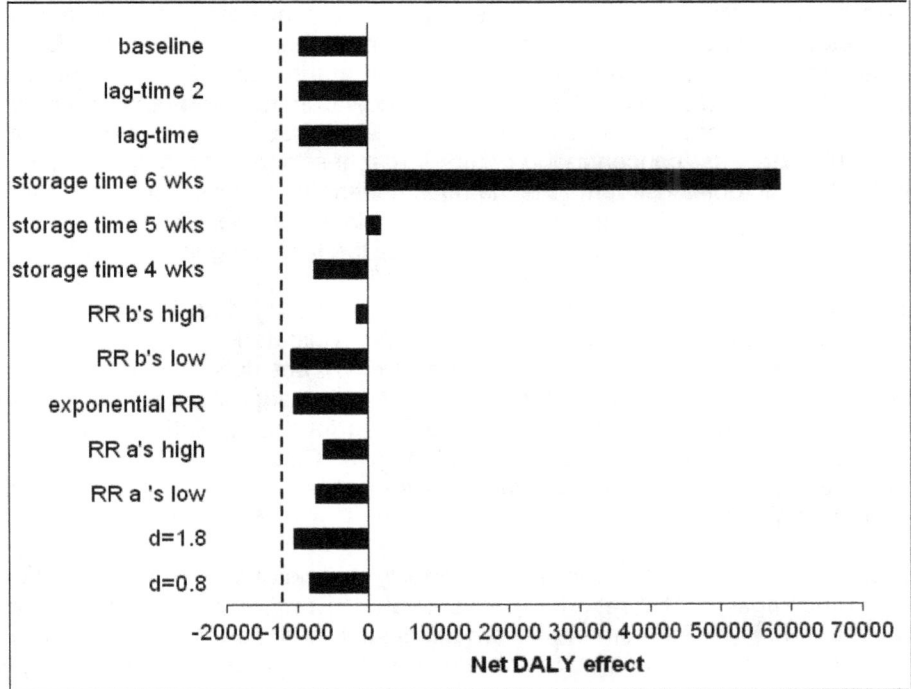

Fig. 2. Result of sensitivity analysis for all selected parameters

When using the exponential function ($\ln(RR) = b* F_{intake}$) the predicted DALY decrease to -5580 and -1964 for CHD mortality and total stroke respectively compared to the DALY estimate obtained by $\ln(RR) = a + b*\ln(F_{intake})$ (Table 7).

4. DISCUSSION

4.1 The CSS Study

In this study with CSS, major health benefit is obtained from the prevention of CHD mortality and the IQ increment of newborns. This is in accordance with other studies on fish consumption and/or omega-3 fatty acid intake (Cohen et al., 2005b; Guevel et al., 2008; Hoekstra et al. 2012). In Hoekstra et al. (2012) the health benefit was higher for stroke than IQ effect, in this study the health benefit is higher for IQ than stroke. This may be related to the difference in dose-response model used for stroke and because salmon is an oily fish that have a significant effect on IQ. As a part of sensitivity analysis the dose-response model used by Hoekstra et al., (2012) for stroke is applied to see the difference and the outcome indicate that the benefit would step up but still the health benefit is higher for IQ gain than for stroke.

Our paper is the first that shows that these health benefits also outweigh the risk of listeriosis, unless the storage time is too long (>4 weeks) and leads to the exposure to high concentrations of the pathogen. Note though that if abortion was valued less severe than in our analysis storage time can be longer before risks outweigh benefits.

As common in microbiological risk assessment, different parameters (storage time, lag-time, growth rate, etc) that affect the concentration of the pathogen have been considered in the exposure assessment of *L. monocytogenes*. However, in the intake assessment of omega-3 fatty acid, the different parameters that affect its concentration have not been considered in the same way. For instance, processing and storage of CSS, and the biovailability of the compound could affect its concentration before it reaches to target organ. In a comparative approach of the microbial risk and the nutritional benefit, the exposure assessment for nutrients and chemical contaminants should consider all the factors that affect the concentration until the compound in question reaches the target organ to exert an action.

The dose-response models that are applied for CHD mortality and stroke used aggregate data and different models have been used and validated to optimize the output. Nevertheless, the conversion of the estimated RRs to absolute risk is associated with uncertainty (Table 1, eq. 5 and 6). On the other hand, a few aggregate data have been used to model the dose-response relationship of the hazardous endpoints. In addition, in this study the intake assessment uses a point estimate of CSS and DHA intake for all age groups but take into account the differences between sexes. These all together might have an impact on the final outcome.

The result is expressed in DALY (morbidity, mortality and recovery) and it appears that, the septicemia to meningitis DALY ratio is approximately 2:1 (Table 7). Compared to men, the DALY changes are higher for women for both septicemia and meningitis cases this could be linked to the increase in intake of CSS on women than men. Looking back the history of invasive listeriosis in Denmark, in most cases men have higher incidence of invasive listeriosis than women (Gerner-Smidt et al. 2005). Septicemia has been the highest morbidity compared to meningitis. On the other hand, the mortality rate is higher for meningitis than septicemia (Gerner-Smidt et al. 2005). However, mostly in comparative studies on overall invasive listeriosis (septicemia vs meningitis), the ratio of septicemia to meningitis is 5:1 (Jensen et al., 2010); 3:1 (Gerner-Smidt et al. 2005).

In this study four parameters have been tested for sensitivity. The most sensitive parameter was the storage time. The shift of a net public health benefit to a net public health risk is observed when CSS is consumed at five weeks of storage and further (Fig. 2). The shift is mainly because of the increased concentration of *L.monocytogenes* that entails increased incidence of listeriosis. According to our model prediction, maximum benefit with minimum risk can be attained from the consumption of CSS within two weeks of production. The risks increase with storage time whereas benefits remain unchanged. Further study may encompass stochastic analysis of all the uncertain parameters presented in the study.

The study was not meant to thoroughly address all beneficial and hazardous components; neither includes the all related endpoints in connection to CSS consumption. Instead, focus was on the integration of microbial hazards and nutritional benefits. Although there could be more endpoints due to the intake/exposure of the selected risk and benefit, this paper assess health outcomes with strong evidences that enable quantitative evaluation. In addition, we evaluated only endpoints that we expected to have relatively high public health impact. For example, febrile gastroenteritis is usually caused by *L. monocytogenes* (WHO/FAO, 2004); but reported quantitative data with respect to this endpoint are insufficient to do a risk-benefit assessment and moreover, febrile gastroenteritis has less public health impact than septicemia and meningitis. Thus, febrile gastroenteritis is not considered in this assessment. Other health effects related to fish/omega-3 fatty acid intake, for instance neuropsychiatric disorders (Young and Conquer, 2005) are not considered for the same reason. If our study would have included chemical hazards as well (dioxin-like compounds and mercury) and would have included all the endpoints, the net DALY would change quantitatively, and the balance between risk and benefit might change as well.

4.2 Implications for Future Risk-Benefit Assessment Methodology

If a pathogen is selected as a hazard, it is essential that the specific associated endpoints are considered instead of considering the generic clinical syndrome. Nevertheless, most Quantitative Microbial Risk Assessments (QMRA) report the generic clinical syndrome cases only. For example, if the assessment includes listeria then the endpoint is cases of listeriosis (Pouillot et al., 2007; Lindqvist and Westoo, 2000; WHO/FAO, 2004). However, these types of data are insufficient if one intends to integrate pathogens in risk-benefit assessment because of the exclusion of the mortality, morbidity, recovery and/or sequela of the specific diseases. In this risk-benefit assessment, the specific major clinical syndromes of foodborne listriosis (meningitis, septicemia and abortion/stillborne) are included, instead of the generic endpoint "cases of listeriosis". On the other hand, this could also be a problem in dose-response modeling as most studies on the pathogenesis of pathogens have dose-response parameters only for general cases like listeriosis, salmonelosis, campylobacteriosis. In this case, different epidemiological data could be used to extrapolate the percentage of the specific endpoints as in section 2.7.

In addition, the integration of the health effects of microbial hazard into risk-benefit assessment may need further refining in the distinction of pathogenesis from the pathogens itself and the microbial toxins especially with regard to exposure assessment and dose-response modeling. Depending on the pathogens/toxins, the exposure assessment and the dose-response relationship require additional investigation like for example the stomach and small intestine dynamics as explained by (Pielaat et al., 2005) for *Bacillus cereus*.

Moreover, in this study the DALY is used as a common health metric to integrate microbial hazard in risk-benefit assessment. Here a thorough quantitative assessment has been done to the end, as opposed to the EFSA, 2010; Hoekstra, et al., 2010) where the assessment stops when the benefit outweighs the risk or vice versa. Had we followed the tiers of BRAFO tiered approach (Hoekstra et al., 2010) in this study, we might have stopped at the earlier stage. For example, in our assessment the baseline result showed that, the benefit clearly outreaches the risk (Table 7); as of Hoekstra, et al., (2010), we could have stopped at this point. However, further quantitative analysis gives a different result in connection with the change in storage time of CSS (Fig. 1), which, in our view, is an important result. This shows that the integration of microbial risk and/or benefit into risk-benefit assessment may require a more elaborate quantitative assessment to reach to best estimation of public health impact.

Furthermore, in general some disease may have some major secondary disease (sequela) that result from the primary clinical syndrome. For instance, if someone gets liseriosis meningitis then there is some probability that person would get neurological sequela (Aouaj et al., 2002). The DALY model applied in this study does not consider this kind of endpoints. However it can be extended to do so easily.

Consequently, the current risk-benefit assessment framework/models need more refinement in regard to the aforementioned points.

5. CONCLUSION

This paper evaluated and integrated the major risk and benefits in connection with the change in CSS consumption. Our risk-benefit assessment predicted that the overall health impact of change in consumption of CSS from reference to alternative intake provide more health benefits for the Danish population.

The model predictions depend on the assumptions taken during the analysis and the sensitivity analysis reveals that the most sensitive parameter was the storage time. If CSS is consumed after two weeks of storage, the benefit remains the same but the risk increases significantly with storage time.

This study provides an insight for future improvement of the methodologies with regard to exposure assessment of the different component, dose-response relationship and common health metric and general framework for risk-benefit assessment.

COMPETING INTERESTS

Authors have declared that no competing interests exist.

REFERENCES

Albert, C.M., Hennekens, C.H., O'Donnell, C.J., Ajani, U.A., Carey, V.J., Willett, W.C., Ruskin, J.N., Joann, E., Manson, J.E. (1998). Fish consumption and risk of sudden cardiac death. JAMA., 279, 23–28.

Allerberger, F., Wagner, M. (2010). Listeriosis: a resurgent foodborne infection. Clin. Micro. Infec., 16, 16–23.

Andersen, K.K., Olsen, T.S., Dehlendorff, C., Kammersgaard, L.P. (2009). Hemorrhagic and ischemic strokes compared: stroke severity, mortality and risk factors. Stroke, 40, 2068-2072.

Aouaj, Y., Spanjaard, L., van Leeuwen, N., Dankert, J. (2002). *Listeria monocytogenes* meningitis: serotype distribution and patient characteristics in the Netherlands, 1976-95. Epidemiol Infect., 128, 405-409.

Ascherio, A., Rimm, E.B., Stampfer, M.J., Giovannucci, E.L., Willett, W.C. (1995). Dietary intake of marine n-3 fatty acids, fish intake, and the risk of coronary disease among men. N. Engl. J. Med., 332, 977-982.

Cohen, J.T., Bellinger, D.C., Connor, W.E., Shaywitz, B.A. (2005a). A quantitative analysis of prenatal intake of n-3 polyunsaturated fatty acids and cognitive development. Am. J. Prev. Med., 29, 366 –374.

Cohen, J.T., Bellinger, D.C., Connor, W.E., Kris-Etherton, P.M., Lawrence, R.S., Savitz, D.A., Shaywitz, B.A., Teutsch, S.M., Gray, G.M. (2005b). A quantitative risk-benefit analysis of changes in population fish consumption. Am. J. Prev. Med., 29, 325–334.

Daviglus, M.L., Stamler, J., Orencia, A.J., Dyer, A.R, Liu, K., Greenland, P., Walsh, M.K., Morris, D., Shekelle, R.B. (1997). Fish consumption and the 30-year risk of fatal myocardial infarction. N. Engl. J. Med., 336, 1046-1053.

Denmark Statistics. (2011). http://www.statbank.dk/statbank5a/default.asp?w=1280

Denmark Technical University. (2011). National Food Institute http://www.foodcomp.dk/v7/fcdb_default.asp.

Ekstrøm, C.L., Sørensen, H. (2011). Introduction to statistical data analysis for the life science. CRC Press, USA.

European Food Safety Authority (EFSA). (2010). Guidance on human health risk-benefit assessment of foods. Scientific opinion. EFSA Journal, 8, 1673.

FAO/WHO. (2011). Report of the Joint FAO/WHO expert consultation on the risks and benefits of fish consumption. Rome, Food and Agriculture Organization of the United Nations; Geneva, World Health Organization.

Gerner-Smidt, P., Ethelberg, S., Schiellerup, P., Christensen, J.J, Engberg, J., Fussing, V., Jensen, A., Jensen, C., Petersen, A.M., Bruun, B.G. (2005). Invasive listeriosis in Denmark 1994–2003: a review of 299 cases with special emphasis on risk factors for mortality. Clin Microbiol, 11, 618–624.

Gillum, R.F., Mussolino, M.E., Madans, J.H. (1996). The relationship between fish consumption and stroke incidence. The NHANES I epidemiologic follow-up study (National Health and Nutrition Examination Survey). Arch Intern Med, 156, 537-542.

Gladyshev, M.I., Sushchik, N.N., Anishchenko, O.V., Makhutova, O.N., Kalachova, G.S., Gribovskaya, I.V. (2008). Benefit-risk ratio of food fish intake as the source of essential fatty acids vs. heavy metals: A case study of Siberian grayling from the Yenisei River. Food Chemistry, 115, 545-550.

Guevel, M.R., Sirot, V., Volatier, J.L., Leblanc, J.C. (2008). A risk-benefit analysis of French high fish consumption: A QALY approach. Risk Analysis, 28, 37-48.

Hansen, L.T., Røntved, S.D., Huss, H.H. (1998). Microbiological quality and shelf life of cold-smoked salmon from three different processing plants. Food Microbiology, 15, 137-150.

Havelaar, A.H., De Hollander, A.E., Teunis, P.F., Evers, E.G., Van Kranen, H.J., Versteegh, J.F., Koten, V., Joke, E.M., Slob, W. (2000). Balancing the risks and benefits of drinking water disinfection: disability adjusted life-years on the scale. Environ. Health Perspect, 108, 315-321.

He, K., Rimm, E.B., Merchant, A., Rosner, B.A., Stampfer, M.J., Willett, W.C., Ascherio, A. (2002). Fish consumption and risk of stroke in men. JAMA, 288, 3130-3136.

Hoekstra, J., Hart, A., Boobis, A., Claupein, E., Cockburn, A., Hunt, A., Knudsen, I., Richardson, D., Schilter, B., Schutte, K., Torgerson, P.R., Verhagen, H., Watzl, B., Chiodini, A. (2010). BRAFO tiered approach for risk-benefit assessment of foods. Food Chem., Toxico. (Article in press).

Hoekstra, J., Hart, A., Owen, H., Zeilmaker, M., Bokkers, B., Thorgilsson, B., Gunnlaugsdottir, H. (2012). Fish, contaminants and human health: Quantifying and weighing benefits and risks. Food Chem. Toxicol. (Article in press).

Hu, F.B., Bronner, L., Willett, W.C., Stampfer, M.J., Rexrode, K.M., Albert, C.M., Hunter, D., Manson, J.E. (2002). Fish and omega-3 fatty acid intake and risk of coronary heart disease in women. JAMA, 287, 1815-1821.

Ingall, T. (2004). Stroke—incidence, mortality, morbidity and risk. J Insur Med., 36, 143–152.

Iso, H., Rexrode, K.M., Stampfer, M.J., Manson, J.E., Colditz, G.A., Speizer, F.E., Hennekens, C.H., Willett, W.C. (2001). Intake of fish and omega-3 fatty acids and risk of stroke in women. JAMA, 285, 304-312.

Jarvinen, R., Knekt, P., Rissanen, H., Reunanen, A. (2006). Intake of fish and long-chain n-3 fatty acids and the risk of coronary heart mortality in men and women. British Journal of Nutrition, 95, 824–829.

Jensen, A.K., Ethelberg, S., Smith, B., Nielse, E.M., Larsson, J.K., Mølbak, J.K., Christensen, J.J., Kemp, M. (2010). Substantial increase in listeriosis, Denmark 2009. Euro Surveill., 15, 1-4.

Jørgensen, L.V., Huss, H.H. (1998). Prevalence and growth of Listeria monocytogenes in naturally contaminated seafood. International Journal of Food Microbiology, 42, 127–131.

Kemmeren, J., Mangen, M-J., van Duynhoven, Y.T., Havelaar, A. (2006). Priority setting of foodborne pathogens. 330080001/2006. National Institute for Public Health and the Environment, the Netherlands (RIVM).

Kromhout, D., Bosschieter, E.B., Coulander, C.L. (1985). The inverse relation between fish consumption and 20-year mortality from coronary heart disease. N. Engl. J. Med., 312, 1205-1209.

Larsson, S.C., Virtamo, J., Wolk, A. (2011). Fish consumption and risk of stroke in Swedish women. Am. J. Clin. Nutr., 93, 487-493.

Lindqvist, R., Westoo, A. (2000). Quantitative risk assessment for *Listeria monocytogenes* in smoked or gravad salmon and rainbow trout in Sweden. Int. J. Fo. Micro., 58, 181–196.

Leroi, F., Joffraud, J.J., Chevalier, F., Cardinal, M. (2001). Research of quality indices for cold-smoked salmon using a stepwise multiple regression of microbiological counts and physico-chemical parameters. J. App. Microbio., 90, 578-587.

Magnússon, S.H., Gunnlaugsdóttir, H., van Loveren, H., Holm, F., Kalogeras, N., Leino, O., Luteijn, J.M., Odekerken, G., Pohjola, M.V., Tijhuis, M.J., Tuomisto, J.T., Ueland, O, White, B.C., Verhagen, H. (2012). State of the art in benefit-risk analysis: Food microbiology. Food. Chem. Toxicol., 50, 33-39.

Mann, J.I., Appleby, P.N., Key, T.J., Thorogood, M. (1997). Dietary determinants of ischaemic heart disease in health conscious individuals. Heart, 78, 450-455.

Mozaffarian, D. (2006). Fish intake, contaminants, and human health: evaluating the risks and the benefits part 1 – health benefits. JAMA, 296, 1885-99.

Mozaffarian, D., Lemaitre, R.N., Kuller, L.H., Burke, G.L., Tracy, R.P., Siscovick, D.S. (2003). Cardiac benefits of fish consumption may depend on the type of fish meal consumed: the cardiovascular health study. Circulation, 107, 1372-1377.

Mozaffarian, D., Longstreth, W.T., Jr, Lemaitre, R.N., Manolio, T.A., Kuller, L.H., Burke, G.L., Siscovick, D.S. (2005). Fish consumption and stroke risk in elderly individuals: the cardiovascular health study. Arch Intern Med, 165, 200-206.

Mylonakis, E., Paliou, M., Hohmann, E.L., Calderwood, S.B., Wing, E.J. (2002). Listeriosis during pregnancy: a case series and review of 222 cases. Medicine (Baltimore), 81, 260-269.

Oomen, C.M., Feskens, E.J., Rasanen, L., Fidanza, F., Nissinen, A.M., Menotti, A., Kok, F.J., Kromhout, D., (2000). Fish consumption and coronary heart disease mortality in Finland, Italy and The Netherlands. Am. J. Epidemiol., 151, 999-1006.

Orencia, A.J., Daviglus, M.L., Dyer, A.R., Shekelle, R.B., Stamler, J. (1996). Fish consumption and stroke in men. 30-year findings of the Chicago Western Electric Study. Stroke, 27, 204-209.

Pielaat, A., Wijnands, L.M., Takumi, K., Nauta, M.J., Van Leusden, F.M. (2005). The fate of *Bacillus cereus* in the gastrointestinal tract. Report, RIVM, Bilthoven, 59.

Pouillot, R., Miconnet, N., Afchain, A.L., Delignette-Muller, M.L., Beaufort, A., Rosso, L., Denis, J.B., Cornu, M. (2007). Quantitative risk assessment of Listeria monocytogenes in French cold-smoked salmon: II. Risk characterization. Society for Risk Analysis, 29, 806-819.

Pedersen, A.N., Fagt, S., Groth, M.V., Christensen, T., Matthiessen, J., Andersen, N.L., Kørup, K., Hartkopp, H., Ygil, H.K., Hinsch, H., Saxholt, E., Trolle, E. (2010). Danskernes kostvaner 2003-2008 Hovedresultater. DTU Fødevareinstituttet, 1-196.

Ross, T (1999). Predictive food microbiology models in the meat industry. Meat and Livestock. Australia, Sydney, Australia, 196.

Saltelli, A., Chan, K., Scott, E.M. (2000). Sensitivity analysis: gauging the worth of scientific models. 1st ed. John Wiley and Sons Chichester, 504.

Smith, B., Kemp, M., Ethelberg, S., Schiellerup P., Bruun, B.G., Gerner-Smidt, P., Christensen, J.J. (2009). *Listeria monocytogenes*: Maternal-foetal infections in Denmark 1994-2005. Scan. J. Infec., 41, 21-25.

Stouthard, M.A.E., Essink-Bot, M.L., Bonsel, G.J., Barendregt, J.J., Kramer, P.G.N., Gunning-Schepers, L.J., van der Maas, P.J. (1997). Department of Public Health Erasmus University Rotterdam.

Wang, M.P., Thomas, G.N., Ho, S.Y., Lai, H.K., Mak, K.H., Lam, T.H. (2011). Fish consumption and mortality in Hong Kong Chinese—the LIMOR study. Ann. Epidemio., 21, 164-169.

WHO. (2008). The global burden of disease: the 2004 update, first ed. WHO, Geneva.

WHO/FAO. (2004). Risk assessment of *Listeria monocytogenes* in ready-to-eat foods. Technical report of microbiological risk assessment series 5, WHO, Geneva.

Young, G., Conquer, J. (2005). Omega-3 fatty acids and neuropsychiatric disorders. Reprod. Nutr. Dev., 45, 1-28.

Zeilmaker, M.J., Hoekstra, J., Jan, C.H., Eijkeren, V., Nynke, J., Hart, A. Kennedy, M., Owen, H. Gunnlaugsdottir, H. (2012). Fish consumption during child bearing age: A quantitative risk–benefit analysis on neurodevelopment. Food Chem. Toxicol. (Article in press).

APPENDIX

1. The Relative Risks of CHD Mortality and Total Stroke

Tables 8 and 9 present the results of studies on the relation between fish intake and the relative risks of CHD mortality and stroke. These results are used for dose response modeling. The most appropriate model is selected after analysis from the five functions mentioned below, based on best fit statistics using statistical computing R-programme.

- $RR = a + b * F_{intake}$
- $RR = a + b * \ln(F_{intake})$
- $\ln(RR) = a + b * F_{intake}$
- $\ln(RR) = b * F_{intake}$
- $\ln(RR) = a + b * \ln(F_{intake})$

Where, RR is relative risk; a and b are parameters and F_{intake}, is fish intake per day (Dose).

Table 8. The studies incorporated in the dose-response modeling of CHD mortality: fish intake (g/d) and relative risk (RR)

Reference	Dose (g/d)	RR (95% CI)	Reference	Dose (g/d)	RR (95%CI)
Kromhout et al., 1985	0	1	Mozaffarian et al., 2003	0	1
	7.5	0.63		7	0.78 (0.47, 1.28)
	22.5	0.56		14	0.77 (0.45, 132)
	37.5	0.36		29	0.53 (0.30, 0.96)
	75	0.39		71	0.47 (0.27, 0.82)
Ascherio et al., 1995	0	1	Mann et al., 1997	0	1
	7	0.74 (0.38, 1.45)		7	1.21(0.62, 2.38)
	18	0.86 (0.5, 1.47)		29	1.23 (0.7, 2.17)
	37	0.71 (0.41, 1.21)	Oomen et al., 2000, The Netherlands	0	1
	69	0.54 (0.29, 1.00		10	1 (0.59, 1.68)
	119	0.77 (0.41, 1.44)		35	1.1 (0.68, 1.79)
Daviglus et al., 1997	0	1	Oomen et al., 2000, Finland	10	1
	9	0.88 (0.63, 1.22)		30	0.97 (0.68, 1.38)
	26	0.84 (0.61, 1.17)		70	1.25 (0.89, 1.76)
	67.5	0.62 (0.4, 0.94)	Jarvinen et al. 2006	4	1
Albert et al., 1998	0	1		12·0	0.91 (0.55, 135)
	7.5	1.18 (0.59, 2.26)		19.8	0.77 (0.48, 1.23)
	21	0.82 (0.45, 1.51)		32	0.68 (0.42, 1.12)
	50	0.91(0.5, 1.66)		70·0	0.59 (0.36, 0.99)
	86	0.81(0.41, 1.61)	Hu et al., 2002	0	1
Oomen et al., 2000, Italy	0	1		7	0.8 (0.56, 1.15)
	10	0.94 (0.55,1.59)		14	0.65 (0.46, 0.91)
	30	0.93 (0.53, 1.63)		43	0.72 (0.48, 1.09)
	70	0.67 (0.33, 1.39)		86	0.55 (0.33, 0.91)

Table 9. The studies incorporated in the dose-response modeling of total stroke: Fish intake (g/d) and relative risk (RR)

Reference	Dose (g/d)	RR (95% CI)	Reference	Dose (g/d)	RR (95% CI)
He et al., 2002	0	1	Mozaffarian et al., 2005)	0	1
	7	0.73 (0.48-1.10)		7	0.88 (0.66–1.17)
	14	0.74 (0.52-1.04		36	0.74 (0.56–0.98)
	43	0.67 (0.46-0.96		86	0.77 (0.56–1.07)
	86	0.83 (0.53-1.29)	Orencia et al., 1996	0	1
Iso et al., 2001	0	1		9	0.98 (0.61, 1.59)
	7	0.93 (0.65-1.34)		26	0.94 (0.59, 1.52)
	14	0.78 (0.55-1.12)		50	1.26 (0.74, 2.16)
	43	0.73 (0.47-1.14	Larsson et al., 2011	0	1
	86	0.48 (0.21-1.06)		17	0.87 (0.75, 1.01)
Gillum et al., 1996	0	1		25	0.92 (0.82, 1.09)
	7	0.78 (0.54, 1.12)		36	0.88 (0.76, 1.02)
	14	0.77 (0.53, 1.13)		64	0.84 (0.71, 0.98)
	60	0.55 (0.32, 0.93)	Wang et al., 2011	7	1
Gillum, et al., 1996	0	1		29	0.7 (0.5, 0.98)
	7	1.27 (0.83, 1.96)		80	0.82 (0.61, 1.11)
	14	1.23 (0.79, 1.91)			
	60	0.85 (0.49, 1.46)			

Table 10. The model outputs for the relative risk of CHD mortality and stroke

Models	CHD	Stroke
$RR = a + b \cdot F_{intake}$	$RR = 0.9 - 0.0028 \cdot F_{intake}$ CI 95%: (0.79, 1.01) and (-0.005, -0.0004) p-value: (<0.001) and (0.023)	$RR = 0.935 - 0.0024 \cdot F_{intake}$ CI 95%: (0.82, 1.05) and (-0.005, 0.0001) p-value: (<0.001) and (0.0593)
$RR = a + b \cdot \ln(F_{intake})$	$RR = 1..11 - 0.0964 \cdot \ln(F_{intake})$ CI 95%: (0.86, 1.36) and (-0.172, -0.021) p-value: (<0.001) and (0.014)	$RR = 1.09 - 0.075 \cdot \ln(F_{intake})$ CI 95%: (0.83, 1.34) and (-0.15, 0.003) p-value: (<0.001) and (0.059)
$\ln(RR) = a + b \cdot F_{intake}$	$\ln(RR) = -0.13 - 0.004 \cdot F_{intake}$ CI 95%: (-0.271, 0.02) and (-0.007, -0.0008) p-value: (0.081) and (0.016)	$\ln(RR) = -0.08 - 0.003 (F_{intake})$ CI 95%: (-0.21, 0.05) and (-0.006, 0.0002) p-value: (0.235, 0.038)
$\ln(RR) = b \cdot F_{intake}$	$\ln(RR) = -0.006 \cdot F_{intake}$ CI 95%: (-0.008, -0.0041) p-value: <0.001	$\ln(RR): -0.0045 \cdot F_{intake}$ CI 95%: (-0.006, -0.003) p-value: <0.001
$\ln(RR) = a + b \cdot \ln(F_{intake})$	**$\ln(RR) = 0.17 - 0.137 \cdot \ln(F_{intake})$** **CI 95%: (-0.16, 0.5) and (-0.24, -0.04)** **p-value:(0.3) and (0.008)**	**$\ln(RR) = 0.113 - 0.094 \cdot \ln(F_{intake})$** **CI 95%: (-0.184, 0.41) and (-0.184, -0.004)** **p-value: (0.44) and (0.04)**

The last model is used to estimate the relative risk for each scenario and endpoints.

4

The Combination of a High-protein Formula Diet and Polyglucosamine Decreases Body Weight and Parameters of Glucose and Lipid Metabolism in Overweight and Obese Men and Women

Janina Willers[1*], Stefanie C. Plötz[1] and Andreas Hahn[1]

[1]Institute of Food Science, Leibniz University Hannover, Hannover, Germany.

ABSTRACT

Aims: To examine the efficacy of a weight loss strategy using a high-protein formula diet in combination with a lipid adsorbing fibre, polyglucosamine, over 12 weeks.

Study Design: Randomised, double-blind, placebo-controlled, parallel group study.

Place and Duration of Study: Institute of Food Science, Unit of Nutrition Physiology and Human Nutrition, Leibniz University of Hannover, Germany, between February 2010 and July 2010.

Methodology: 120 overweight and obese subjects participated in this study and ingested a protein-rich formula diet as a meal replacement once daily. Half of the participants (n = 60) additionally took two lipid-adsorbent tablets (F+LA), polyglucosamine, once daily, while the other half (n = 60) received two placebo tablets (F+P). Measurements were taken at weeks 0, 6 and 12 to determine the response to intervention.

Results: Both groups achieved a highly significant (P < 0.001) weight loss (F+LA: -5.5 ± 3.8 kg vs. F+P: -4.7 ± 3.9 kg, Full Analysis Set (FAS) population). There was a significant decrease in HbA1c (P < 0.01), total cholesterol (P < 0.001), LDL cholesterol (P = 0.002), and triacylglycerol (P = 0.001) in the F+LA group, while the F+P group experienced no changes.

Conclusion: The investigation demonstrated that a formula diet alone or combined with polyglucosamine were both effective in weight reduction. The additional administration of polyglucosamine was more effective on the reduction of glucose and lipid parameters than the formula diet alone.

*Corresponding author: Email: willers@nutrition.uni-hannover.de

Keywords: High-protein formula diet; polyglucosamine; obesity; weight reduction; lipid profile.

1. INTRODUCTION

Cardiovascular diseases have become one of the most common and significant health issues of our time. Within the framework of the pathogenesis of cardiovascular diseases, various causes are thoroughly discussed in the literature in which obesity plays a special key role as a dependent and an independent risk factor. In this regard, (abdominal) obesity is considered a prerequisite and a central component of the metabolic syndrome which is characterized by the coexistence of key cardiovascular disease risk factors including obesity, hyperglycemia, hypertension and dyslipidemia (Kloting et al., 2007).

Overweight and obesity in children and adults is a major problem affecting a very significant proportion of populations worldwide (Mensink et al., 2005; Gellner and Domschke, 2008; Apfelbacher et al., 2008; Moura and Claro, 2011). Therefore, an integral part of the strategy with regard to the preventive and therapeutic measures against diseases of the cardiovascular system and even in the presence of metabolic syndrome, is the reduction of body weight, particularly visceral fat mass (Case et al., 2002).

Different nutrition and lifestyle-related treatment methods are currently available for stabilisation of weight and weight loss (Shaw et al., 2006; Brunner et al., 2007; Rolland et al., 2009). Meal replacement strategies (partial and full meal replacements) using formula diets are widely used to attain initial weight loss (Keogh and Clifton, 2005; Carbajo et al., 2010). This has already been confirmed as a health claim by the European Food Safety Authority (EFSA, 2011a).

Due to its easy, customer-friendly features and comparatively low costs, as well as the predetermined serving size, formula diet products are especially suitable for people who have difficulty losing weight. Apart from weight loss, various different studies have shown improvements in metabolic risk factors (glucose, insulin, lipid profile including blood pressure and waist circumference) when using formula diet products as partial meal replacements (Ditschuneit et al., 1999; Flechtner-Mors et al., 2000; Ditschuneit and Flechtner-Mors, 2001; Ashley et al., 2001a, 2001b; Rolland et al., 2009; Smith et al., 2010).

This concept has proved efficacious even in overweight patients with type 2 diabetes (Yip et al., 2001; Redmon et al., 2005; Hamdy and Zwiefelhofer, 2010). In this context, several studies have proved the short-term effectiveness (e.g. 12 weeks) of a meal replacement strategy using formula diet products (Rothacker et al., 2001). Moreover, other studies also showed that, together with long-term follow-ups (e.g. 1 year), formula diets may support weight stabilisation (Rothacker et al., 2001; Noakes et al., 2004).

Lipid adsorbing substances, such as polyglucosamine, can be considered as a further supplementary strategy to support weight loss via increased excretion of faucal fat. Polyglucosamine is derived from the shells of crustaceans, such as prawns, lobsters and crabs. Polyglucosamine fibres have fat adsorbing properties and are, therefore, able to bind therapeutically relevant amounts of fat (Rodriguez and Albertengo, 2005; Yao and Chiang, 2006). Polyglucosamine is protonated in the stomach. In this state, polyglucosamine can bind monomeric fatty acids and bile acids with high affinity in the duodenum by means of electrostatic attraction (the positively charged amino groups of the glucosamine residues (cations) bind the negatively charged carboxyl groups of fatty acids and bile acids (anions)), forming a molecular gel structure (Kanauchi et al., 1995). The fats mainly bound to polyglucosamine can no longer be resorbed and are excreted in the stool.

The absorption of lipids through the intestinal mucosal cells proceeds at a much slower rate in the presence of polyglucosamine, as the bile acids, including the mono fatty acids, the nonpolar lipids and undigested fats (lipophilic interactions), needed for the emulsification are primarily bound to polyglucosamine (Kaats et al., 2006). Due to this mechanism, the use of polyglucosamine can be theoretically useful for weight loss and weight control, as well as for the treatment of hypercholesterolemia (Pittler et al., 1999; Schiller et al., 2001; Ylitalo et al., 2002; Mhurchu et al., 2004, 2005; Kaats et al., 2006; Jull et al., 2008; Cornelli et al., 2008).

Within this context, the treatment effect of a weight loss programme using a high-protein formula diet (formoline protein diet) in combination with a lipid adsorbing (LA) fibre, polyglucosamine L112, compared with the treatment group without polyglucosamine L112 was investigated in this parallel-group trial. The dose of polyglucosamine L112 (0.8 g/d) was taken according to weight stabilization. Hence, we put forward the hypothesis with regard to weight loss that the strategy combining (F+LA) was more effective than the combination, formula diet with placebo (F+P). Changes in different laboratory parameters (lipid profile, carbohydrate metabolism) were additionally assessed in the overweight and obese participants.

2. MATERIALS AND METHODS

2.1 Study Design and Population

This clinical trial was designed as a 12-week, single centre, randomised, placebo-controlled, double-blind, and parallel group study. The trial protocol was developed with methods according to the guidelines for Good Clinical Practice (GCP) and approval of the Freiburg Ethics Commission International (FEKI) was received on 30th November 2009. Written informed consent was obtained from all participants. The clinical investigation was registered in the German Clinical Trials Register (DRKS) with the identification number DRKS00000325.

The study was carried out at the Institute of Food Science, Leibniz University Hannover, Germany. One hundred and twenty overweight (16.7%) and obese (83.3%) participants (61 males and 59 females, all of them Caucasian) were enrolled in the trial in February 2010. The trial participants were recruited through advertisements in the daily newspapers. Study criteria were assessed in advance via structured questionnaires over the telephone. Inclusion criteria for this study were a body mass index (BMI) between 28 and 35 kg/m², age between 30 and 60 years. Major chronic diseases (e.g. cancer diseases, manifest cardiovascular disease (CVD), insulin-dependent type 1 or 2 diabetes, severe renal or liver diseases, endocrine and autoimmune diseases), gastrointestinal disorders (e.g. ulcers, chronic inflammatory bowel diseases, coeliac disease, pancreatitis), prior gastrointestinal surgical procedures (e.g. gastrectomy, short bowel syndrome, gastric bypass, gastric banding, stomach balloon), lactose intolerance, pregnancy, breastfeeding, and alcohol or drug addiction were defined as exclusion criteria. Participants were excluded if they were currently following a diet or taking any supplements that could interfere with the preparations given.

After returning the completed admission questionnaire sent by post, participants who met the inclusion criteria without the presence of any of the exclusion criteria were included and invited to the first examination. Patients (stratified according to their gender) were assigned to their respective intervention groups by means of block randomisation appropriate to the sample size, sorted in ascending order of the patient numbers. The chief investigator, investigators, study staff, and patients were all blinded to the treatment allocation in accordance with the double-blind design.

All 120 participants consumed one serving of protein-rich formula diet (formoline protein diet, Certmedica International GmbH, Aschaffenburg, Germany) a day (either lunch or dinner). It was administered daily as a drink replacement meal. The formula diet was made by mixing three tablespoons (approximately 23 g) of powder plus 300 ml milk (low fat) plus 5 g vegetable oil together. The nutritional composition of the formula diet is presented in Table 1.

Table 1. Nutrient and energy content of the formula diet with recommended preparation*

	Preparation per serving	Preparation per 100 ml
Energy [kJ]	1319 (313 kcal)	389 (92 kcal)
Protein [g]	29.3	8.6
Carbohydrate [g]	25.6	7.6
Fat [g]	10.2	3.0
Vitamin A [µg]	339	100
Vitamin D [µg]	1.55	0.5
Vitamin E [mg]	7.2	2.1
Vitamin C [mg]	30.1	8.8
Vitamin B1 [mg]	0.7	0.2
Vitamin B2 [mg]	2.1	0.6
Niacin [mg]	7.8	2.3
Vitamin B6 [mg]	1.0	0.3
Folate [µg]	131.0	38.6
Vitamin B12 [µg]	3.3	1.0
Biotin [mg]	0.04	0.01
Pantothenic acid [mg]	3.6	1.1
Calcium [mg]	604.0	178.2
Phosphor [mg]	384.0	113.3
Iron [mg]	4.9	1.4
Zinc [mg]	4.0	1.2
Copper [mg]	0.34	0.1
Iodine [µg]	50.0	14.7
Sodium [g]	0.301	0.089
Magnesium [mg]	92.0	27.1
Manganese [mg]	0.3	0.1
Potassium [mg]	530.0	156.3
Selenium [µg]	20.0	5.9

*Recommended preparation: 300 ml milk (low fat) + 23 g powder + 5 g vegetable oil

In addition to the formula diet, participants in group F+LA (n = 60) received two lipid adsorbent tablets (polyglucosamine L112, Certmedica International GmbH, Aschaffenburg, Germany), which were taken once a day with the main meal of the day. In the specification of polyglucosamine L112, the main active ingredient in the lipid adsorbing verum preparation is a natural, indigestible fibrous material (β-1,4 polymer of D-glucosamine and N-acetyl-D-glucosamine). One tablet of the verum preparation contains approximately 400 mg of the active ingredient, polyglucosamine. The intake of two tablets per day is the manufacturer's recommended amount for weight maintenance. Further excipients are cellulose from plants, vitamin C, tartaric acid, silicon dioxide, and magnesium stearate (vegetable origin). The preparation does not contain flavour enhancers, colours, preservatives, gelatine, gluten, lactose, iodine, or cholesterol.

In addition to the formula diet, the participants in group F+P were given two placebo tablets which mainly contained cellulose (245 mg/2 tablets), calcium hydrogen phosphate (745 mg/2 tablets) and magnesium stearate (10 mg/2 tablets). There were no apparent differences in colour, size, shape, smell, and taste between the LA and placebo tablets. The participants were instructed to replace either lunch or dinner (depending on the daily routine) with the protein-rich formula diet. They also received recipe suggestions for the meal replacement drinks to have variation in the preparation. Breakfast and every other meal were prepared according to the principles of a balanced wholefood diet and based on national recommendations (German Nutrition Society). A recipe book adapted to suit the clinical trial was given to all the study participants to encourage healthy eating habits. Two tablets of either LA or P should be taken together with the correspondingly other main meal of the day (which could be either lunch or dinner). The required amount of investigational products in neutral packaging was provided to the study participants prior to the trial commencement and after six weeks. At the end of the study, the remaining tablets were counted to check compliance. Participants also received two sessions of nutrition education and six sessions of physical activity training as an adjunct to the intervention. Furthermore, all the study participants were given elastic bands (Therabands) including exercise handouts (information sheets and workout DVDs) to promote and support physical activity at home.

2.2 Questionnaires, Anthropometric Measurements and Body Composition

Anthropometric measurements were taken at each time point, in particular at baseline (week 0) and after 6 and 12 weeks (t_0, t_6, t_{12}). Height was measured using a stadiometer (SECA, Model 217, Seca Gmbh & Co Kg, Germany). Weight was measured with the participants dressed in light clothes and without shoes. An amount of 1 kg was deducted from the weight measured as an allowance for the clothing. Waist circumference (WC) was measured at the midpoint between the lower border of the rib cage and the top of the iliac crest. Hip circumference (HC) was defined as being the widest circumference over the buttocks. The measurements of the WC and HC were taken with the participants standing relaxed, breathing normally, with the measuring tape placed horizontally. Body composition was determined by means of bioelectrical impedance analysis (BIA) using a calibrated device, the Nutriguard-M (Data Input GmbH, Darmstadt, Germany).

2.3 Blood Sampling and Laboratory Procedures

Fasting blood samples (approximately 45 ml at each time point week 0, 6, 12) were collected using sealed Blood Collection Tubes and System S-Monovettes® (Sarstedt, Germany). The blood samples were centrifuged at 2000 x g for 10 minutes immediately after collection to prepare the plasma. Blood samples were allowed to clot for 20-30 minutes in the refrigerator and were subsequently centrifuged under the above-mentioned conditions to separate the serum. The serum triacylglycerol (TAG) concentration, total cholesterol (TC) and HDL cholesterol (HDL-C) levels were measured enzymatically (Beckman Coulter, Inc.). LDL cholesterol (LDL-C) was calculated using the Friedewald equation. HbA1c was measured in EDTA whole blood using ion exchange chromatography (Bio-Rad Laboratories, Germany), which is a method of high-performance liquid chromatography (HPLC). Fasting plasma glucose measurements were performed using the hexokinase method, an ultraviolet (UV) enzymatic (*in vitro*) assay (Beckman Coulter, Inc.). Serum insulin concentrations were determined by immunoassays (cobas®, Roche Diagnostics, Mannheim, Germany).

2.4 Detection of Cardiovascular Risk Factors

The systolic (SBP) and diastolic blood pressures (DBP) were measured at rest using a digital device, Tensoval (Paul Hartmann AG, Heidenheim, Germany). Readings were taken three times at an interval of three minutes and at each time point (week 0, 6, 12). The mean value was calculated from this set of readings. Those whose blood pressure was ≥ 130 mmHg (SBP) and/or ≥ 85 mmHg (DBP) were defined as hypertensive. Those study participants with lower blood pressure levels and already diagnosed with hypertension and/or were taking antihypertensive medications (n = 22) were also classified as hypertensive. In total, 104 (86.7%) participants were hypertensive (n = 53 male, n = 51 female). The criteria used in the present study for defining the metabolic syndrome (MetS) according to Alberti et al. (2009) are shown in Table 2.

Table 2. Criteria for identification of the metabolic syndrome, defined by at least three of the five risk factors from Alberti et al. (2009)

Risk factor	Defining level
Elevated waist circumference	Men ≥ 102 cm
	Women ≥ 88 cm
TAG	≥ 150 mg/dl (1.7 mmol/l)
HDL-C	Men < 40 mg/dl (1.0 mmol/l)
	Women < 50 mg/dl (1.3 mmol/l)
Fasting glucose	≥ 100 mg/dl
Blood pressure	Systolic ≥ 130 and/or diastolic ≥ 85 mmHg

TAG: triacylglycerol; HDL-C: HDL cholesterol

2.5 Statistical Analysis

The statistical data analysis was carried out by using the Statistical Package for Social Sciences SPSS 19.0 (SPSS Inc., Chicago, Illinois, USA). The results were shown as the mean value ± standard deviation (s.d.). Differences between groups F+LA and F+P were calculated using the Mann-Whitney U test. The changes in the parameters in comparison with baseline were analysed using the Wilcoxon test. The chi-square test was used to compare the difference between the frequencies of the two groups. P values < 0.05 were interpreted as statistically significant. For data analysis within the framework of the population analysed, missing data were replaced by the last available values, provided at least one data value and the baseline value (t_0) could be obtained.

The Full Analysis Set (FAS) population (n = 106) comprised those participants with at least a baseline value (t_0) and at least one further measurement value. All 120 randomised participants were assessed on the basis of the Intention-To-Treat (ITT) analysis. Missing values were constantly updated in participants without any further values after the baseline measurements.

3. RESULTS

3.1 Baseline Characteristics

The data sets collected within the framework of this intervention study were obtained from 120 participants who were included in the study during the time period between February 15 and 26, 2010 (basic analysis). Subjects from the waiting list moved up if others did not fulfil the BMI criteria at the baseline visit. Sixty participants were randomised to group F+LA and another 60 participants to group F+P (ITT-Population). Fourteen participants (11.7%) who dropped out of the study before week 6 (t_6) because of intolerance to the study medication (n = 3), acute illnesses (n = 7) or other reasons were (n = 4) were excluded from the FAS evaluation.

Consequently, 106 participants were included in the FAS analysis: 52 of them received LA tablets (n = 27 male, n = 25 female) and 54, P tablets (n = 27 male, n = 27 female). A flowchart is depicted in Figure 1.

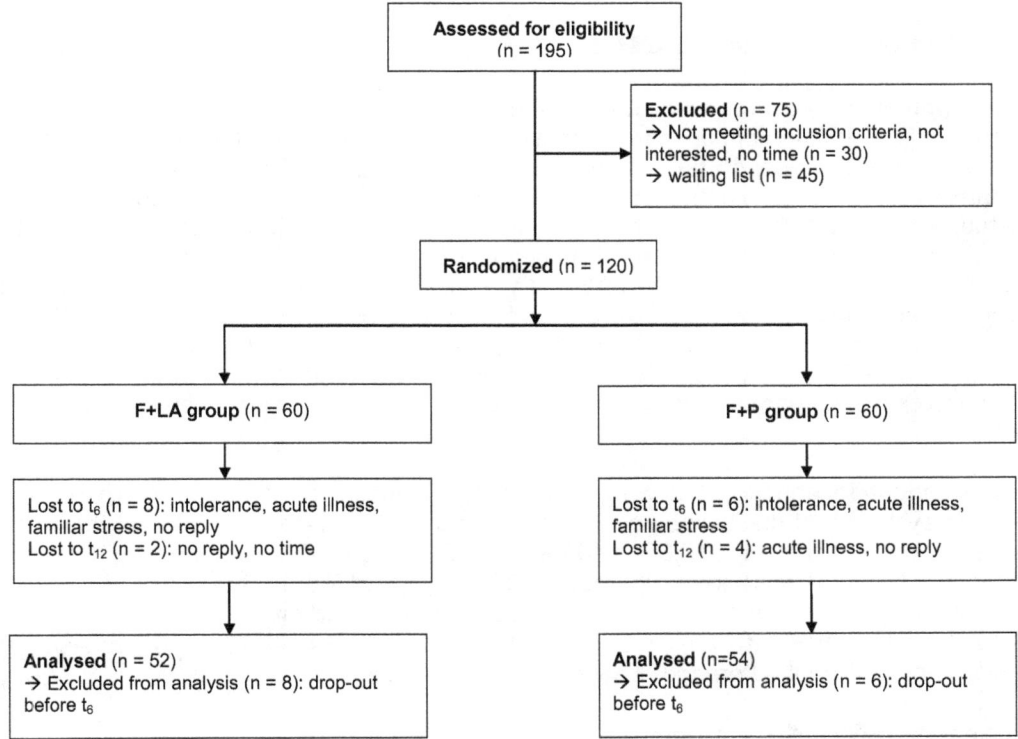

Figure 1. Flowchart
F+LA: formula diet + lipid-adsorbent tablets; F+P: formula diet + placebo tablets

There were no statistically significant differences between groups F+LA (n = 52) and F+P (n = 54) at the beginning of the trial with respect to the reported baseline parameters listed in Table 3. With trend significance, the F+P group has a higher age and DBP compared to the F+LA group. In view of the blood parameters, there were no significant differences between the groups at baseline (Table 5).

Table 3. Baseline participant's characteristics (full analysis set)

	Total group	F+LA group	F+P group	P*
Number of participants [n]	106	52	54	
Sex distribution [m/f]	54 / 52	27 / 25	27 / 27	0.843**
Age [years]	46.9 ± 7.2	45.4 ± 7.0	48.3 ± 7.1	0.061
Initial weight [kg]	97.0 ± 11.7	96.8 ± 12.8	97.1 ± 10.7	0.865
Height [m]	174.7 ± 9.8	174.5 ± 9.5	175.0 ± 10.1	0.884
BMI [kg/m²]	31.7 ± 1.95	31.7 ± 2.02	31.7 ± 1.9	0.987
WC [cm]	106.1 ± 8.1	105.6 ± 7.8	106.5 ± 8.3	0.702
HC [cm]	114.6 ± 4.8	114.1 ± 4.8	115.0 ± 4.9	0.390
SBP [mmHg]	138.5 ± 15.7	135.8 ± 13.1	141.0 ± 17.5	0.227
DBP [mmHg]	93.3 ± 11.0	90.9 ± 9.1	95.6 ± 12.3	0.060
MetS [%]	63.2 (n = 67)	63.5 (n = 33)	63.0 (n = 34)	

*Mann-Whitney U test; **chi-square test; BMI: body mass index; WC: waist circumference; HC: hip circumference; SBP: systolic blood pressure; DBP: diastolic blood pressure; MetS: prevalence of metabolic syndrome; F+LA: formula diet + lipid-adsorbent tablets; F+P: formula diet + placebo tablets*

3.2 Changes in Anthropometric Data

The anthropometric measurements observed in both the FAS and ITT populations for the two groups after an intervention period of twelve weeks confirmed the highly significant changes ($P < 0.001$; Wilcoxon test) (Table 4). Weight loss in the F+LA group was 0.74 kg more than in the F+P group, although the difference between the two groups turned out not to be statistically significant. Men decreased significantly more weight than women (-5.3 ± 4.5 kg vs. 3.6 ± 3.4 kg, $P < 0.05$; Mann-Whitney U test). Altogether, 95% of the participants in the allocated group (n = 101) in the FAS lost weight during the study period, 41% (n = 43) achieved a weight loss of around 5 kg or more. The proportion of about 46% (n = 24) in group F+LA was higher than in group F+P (35%, n = 19). However, this difference was not significant ($P = 0.283$; chi-square test). There were trends for greater improvements in further anthropometric parameters (e.g. BMI, WC) in the F+LA group relative to the F+P group.

After an intervention period of 12 weeks, a highly significant ($P < 0.001$; Wilcoxon test) reduction in systolic and diastolic blood pressures and in heart rate were seen in the whole study population (SBP Δ t_{12}-t_0: -10.9 ± 11.0 mmHg; DBP Δ t_{12}-t_0: -6.6 ± 7.0 mmHg; heart rate Δ t_{12}-t_0: -5.7 ± 7.9 beats/min): in group F+LA (SBP Δ t_{12}-t_0: -11.7 ± 10.0 mmHg; DBP Δ t_{12}-t_0: -5.7 ± 6.8 mmHg; heart rate Δ t_{12}-t_0: -4.5 ± 7.2 beats/min) and in group F+P (SBP Δ t_{12}-t_0: -10.2 ± 12.0 mmHg; DBP Δ t_{12}-t_0: -7.6 ± 7.2 mmHg; heart rate Δ t_{12}-t_0: -6.7 ± 8.5 beats/min). There were no significant differences between the two groups.

3.3 Changes in Blood Parameters

The HbA1c levels ($P < 0.01$; Wilcoxon test), TC ($P < 0.001$; Wilcoxon test), LDL-C ($P = 0.001$; Wilcoxon test), and TAG levels ($P = 0.001$; Wilcoxon test) in group F+LA, as well as the FAS and ITT populations, were significantly reduced compared to baseline levels, while no significant changes were experienced in group F+P. When the intervention phase ended, group F+LA demonstrated significantly lower TC and LDL-C levels than group F+P ($P < 0.01$; Wilcoxon test). A significant reduction in insulin levels of around 30% ($P < 0.001$; Wilcoxon test) was measured in both groups. The changes in the measurement of biochemical parameters are shown in Table 5.

The percentage of study subjects with elevated TC levels ($\geq$ 6.2 mmol/l) in the FAS population of group F+LA at the time of study entry, dropped from 53% (n = 26) to 25% (n = 12) during the course of the study, whereas the proportion in group F+P rose slightly ($\angle$8%; n = 24 vs. 54%; n = 27, respectively). Similar changes were observed in the concentrations of LDL-C and TAG. The percentage of participants in group F+LA with elevated LDL-C ($\geq$ 4.1 mmol/l) and TAG concentrations ($\geq$ 1.7 mmol/l) dropped from 35% (n = 16) to 24% (n = 11) and from 49% (n = 24) to 35% (n = 17), respectively. No significant change was observed in the elevated LDL-C levels in group F+P. The percentage of participants with an elevated TAG level ($\geq$ 1.7 mmol/l) in group F+P dropped from initially 52% (n = 26) to 44% (n = 22).

With regard to the metabolic syndrome (MetS, at least three out of five criteria), a reduction in the prevalence could be achieved in the whole study population (n = 106). The prevalence in group F+LA fell from 64% (n = 33) to 35% (n = 18) and the incidence could therefore be halved. The prevalence of MetS in group F+P decreased from an initial 63% (n = 34) to 44% (n = 24). The difference between the frequency in the two groups was not statistically significant (P = 0.301; chi-square test).

3.4 Adverse Events and Side Effects

Incidences of all kinds of adverse events (e.g. common cold) in both treatment groups were comparatively similar. However, there were no statistically significant differences between groups F+LA and F+P in the proportion of participants reporting adverse events (F+LA: 15 out of 60; 25% vs. F+P: 12 out of 60; 20%; P = 0.512; chi-square test).

At week 6 there were statistically significant differences between the F+LA and the F+P group in the proportion of participants with flatulence, diarrhea and constipation. While flatulence (n = 4) and diarrhea (n = 4) only occurred in the F+P group (P < 0.05; chi-square test), complaints of constipation were significantly more frequent in the F+LA group (n = 10 vs. n = 3; P = 0.032; chi-square test). Multiple answers were possible here. At the visit t_{-2} no statistically significant differences were seen between the F+LA und F+P group. Effects such as bloating, nausea or eructation were rarely mentioned.

Table 4. Changes of anthropometric data in the FAS and ITT population after 12 weeks of intervention

		FAS					ITT			
	Participants	t_0 mean ± SD	t_{12} mean ±SD	Δ t_{12}-t_0 mean ± SD	$P^{\#}$	Participants	t_0 mean ± SD	t_{12} mean ±SD	Δ t_{12}-t_0 mean ± SD	$P^{\#}$
Weight [kg]	F+LA [n = 52]	96.8 ± 12.8	91.4 ± 12.2	-5.46 ± 3.83	< 0.001	F+LA [n = 60]	96.8 ± 12.2	92.1 ± 11.8	-4.73 ± 4.03	< 0.001
	F+P [n = 54]	97.1 ± 10.7	92.3 ± 10.1	-4.72 ± 3.88	< 0.001	F+P [n = 60]	96.9 ± 10.6	92.6 ± 10.1	-4.24 ± 3.94	< 0.001
P*		0.865	0.702	0.125			0.929	0.912	0.362	
BMI [kg/m²]	F+LA [n = 52]	31.7 ± 2.0	29.9 ± 2.2	-1.76 ± 1.20	< 0.001	F+LA [n = 60]	31.8 ± 2.0	30.2 ± 2.3	-1.53 ± 1.27	< 0.001
	F+P [n = 54]	31.7 ± 1.9	30.2 ± 2.2	-1.50 ± 1.17	< 0.001	F+P [n = 60]	31.9 ± 1.9	30.5 ± 2.3	-1.44 ± 1.23	< 0.001
P*		0.987	0.422	0.159			0.753	0.391	0.376	
WC [cm]	F+LA [n = 52]	105.6 ± 7.8	99.8 ± 7.3	-5.83 ± 3.34	< 0.001	F+LA [n = 60]	105.9 ± 7.8	100.9 ± 7.8	-5.05 ± 3.69	< 0.001
	F+P [n = 54]	106.5 ± 8.3	100.9 ± 8.5	-5.59 ± 3.82	< 0.001	F+P [n = 60]	106.3 ± 8.2	101.2 ± 8.4	-5.03 ± 4.00	< 0.001
P*		0.702	0.588	0.516			0.854	0.823	0.796	
HC [cm]	F+LA [n = 52]	114.1 ± 4.8	109.5 ± 4.8	-4.56 ± 3.32	< 0.001	F+LA [n = 60]	114.1 ± 5.0	110.2 ± 5.2	-3.95 ± 3.46	< 0.001
	F+P [n = 54]	115.0 ± 4.9	111.0 ± 5.0	-3.98 ± 3.32	< 0.001	F+P [n = 60]	115.2 ± 5.1	111.6 ± 5.5	-3.58 ± 3.37	< 0.001
P*		0.390	0.179	0.255			0.281	0.183	0.491	
WHR [WC/HC]	F+LA [n = 52]	0.93 ± 0.07	0.91 ± 0.06	-0.02 ± 0.03	< 0.001	F+LA [n = 60]	0.93 ± 0.07	0.92 ± 0.06	-0.01 ± 0.03	< 0.001
	F+P [n = 54]	0.93 ± 0.08	0.91 ± 0.07	-0.02 ± 0.03	< 0.001	F+P [n = 60]	0.92 ± 0.08	0.91 ± 0.07	-0.02 ± 0.03	< 0.001
P*		0.783	0.830	0.769			0.805	0.707	0.619	

*Mann-Whitney U test; #Wilcoxon test; FAS: Full Analysis Set (n = 106); ITT: intention-to-treat (n = 120): BMI: body mass index; WC: waist circumference; HC: hip circumference; WHR: waist-to-hip ratio; F+LA: formula diet + lipid-adsorbent tablets; F+P: formula diet + placebo tablets

Table 5. Changes of blood parameters in the FAS and ITT population after 12 weeks of intervention

| | FAS | | | | | ITT | | | | |
	Participants	t0 mean ± SD	t12 mean ± SD	Δ t12-t0 mean ± SD	P#	Participants	t0 mean ± SD	t12 mean ± SD	Δ t12-t0 mean ± SD	P#
Glucose [mmol/l]	F+LA [n = 52]	5.16 ± 0.47	5.09 ± 0.55	-0.07 ± 0.43	0.109	F+LA [n = 60]	5.21 ± 0.50	5.15 ± 0.58	-0.06 ± 0.40	0.109
	F+P [n = 54]	5.48 ± 0.85	5.54 ± 1.80	0.06 ± 1.26	0.264	F+P [n = 60]	5.43 ± 0.83	5.48 ± 1.71	0.06 ± 1.19	0.264
P*		0.115	0.084	0.769			0.442	0.354	0.722	
HbA1c [%]	F+LA [n = 52]	5.54 ± 0.25	5.46 ± 0.20	-0.08 ± 0.18	**0.006**	F+LA [n = 60]	5.57 ± 0.28	5.50 ± 0.25	-0.07 ± 0.17	**0.006**
	F+P [n = 53]	5.68 ± 0.48	5.67 ± 0.68	-0.004 ± 0.34	0.398	F+P [n = 59]	5.67 ± 0.47	5.68 ± 0.65	-0.003 ± 0.32	0.398
P*		0.223	**0.027**	0.222			0.341	0.061	0.251	
Insulin [pmol/l]	F+LA [n = 52]	95.5 ± 38.1	68.1 ± 35.6	-27.4 ± 29.9	**< 0.001**	F+LA [n = 60]	96.7 ± 36.3	72.9 ± 36.1	-23.7 ± 29.3	**< 0.001**
	F+P [n = 54]	106.2 ± 61.3	74.6 ± 41.3	-31.6 ± 42.6	**< 0.001**	F+P [n = 60]	108.7 ± 60.6	80.2 ± 45.4	-28.5 ± 41.5	**< 0.001**
P*		0.517	0.509	0.820			0.493	0.558	0.644	
TC [mmol/l]	F+LA [n = 52]	6.12 ± 1.00	5.66 ± 0.82	-0.45 ± 0.75	**< 0.001**	F+LA [n = 60]	6.09 ± 0.97	5.69 ± 0.80	-0.39 ± 0.72	**< 0.001**
	F+P [n = 54]	6.43 ± 1.27	6.39 ± 1.34	-0.04 ± 0.79	0.783	F+P [n = 60]	6.37 ± 1.28	6.33 ± 1.34	-0.04 ± 0.74	0.783
P*		0.236	**0.007**	**0.011**			0.301	**0.017**	**0.021**	
HDL-C [mmol/l]	F+LA [n = 52]	1.40 ± 0.32	1.37 ± 0.30	-0.03 ± 0.15	0.050	F+LA [n = 60]	1.39 ± 0.31	1.36 ± 0.29	-0.03 ± 0.14	0.050
	F+P [n = 54]	1.42 ± 0.32	1.42 ± 0.29	0.00 ± 0.21	0.936	F+P [n = 60]	1.43 ± 0.32	1.43 ± 0.29	0.00 ± 0.20	0.936
P*		0.578	0.359	0.172			0.303	0.177	0.161	
LDL-C [mmol/l]	F+LA [n = 50]	3.86 ± 0.87	3.56 ± 0.73	-0.30 ± 0.60	**0.002**	F+LA [n = 58]	3.85 ± 0.83	3.59 ± 0.71	-0.26 ± 0.56	**0.003**
	F+P [n = 51]	4.13 ± 1.02	4.14 ± 1.07	0.01 ± 0.60	0.866	F+P [n = 57]	4.07 ± 1.05	4.08 ± 1.10	0.01 ± 0.56	0.866
P*		0.216	**0.047**	**0.013**			0.339	**0.023**	**0.022**	
TAG [mmol/l]	F+LA [n = 52]	1.95 ± 1.13	1.68 ± 1.19	-0.27 ± 0.78	**0.001**	F+LA [n = 60]	1.92 ± 1.09	1.69 ± 1.14	-0.24 ± 0.73	**0.001**
	F+P [n = 54]	2.10 ± 1.59	1.86 ± 1.09	-0.23 ± 1.32	0.167	F+P [n = 60]	2.05 ± 1.55	1.84 ± 1.09	-0.21 ± 1.25	0.167
P*		0.719	0.084	0.076			0.908	0.150	0.127	

*Mann-Whitney U test; #Wilcoxon test; FAS: Full Analysis Set (n = 106); ITT: Intention-to-treat (n = 120); TC: total cholesterol; HDL-C: HDL cholesterol; LDL-C: LDL cholesterol; TAG: triacylglycerol; F+LA: formula diet + lipid-adsorbent tablets; F+P: formula diet + placebo tablets

4. DISCUSSION

The study results show that substantial weight loss, on average about 5 kg, can be achieved within twelve weeks by using the investigational product (formula diet) plus a lifestyle and nutritional programme. The administration of the lipid-adsorbent tablets (0.8 g polyglucosamine/day) with the concurrent use of a protein-rich formula diet (F+LA) led to significant changes in biochemical lipid and glucose parameters compared with placebo. There were also significant improvements in the parameters of the BMI, waist circumference, hip circumference, and blood pressure of both groups during intervention. However, the changes were not significantly different between the groups. The changes were also more favourable here for group F+LA, however without significant differences between the groups. The evaluations of all the results from the FAS population were principally confirmed by the more conservative approach of ITT.

Even though men had a significantly higher weight loss than women, the overall weight loss in both groups was comparable. This mainly arises from the stratification by gender which led to an equal distribution.

Both the usage of the formula diet on its own or taken in combination with polyglucosamine were effective for weight loss within the framework of this clinical investigation. It is probable that an increased intake of polyglucosamine L112, 1 x 2 tablets initially, then 2 x 2 tablets a day, may result in a more pronounced effect compared with placebo. It should be noted, however, that a higher dose could also cause more side effects such as constipation, mild nausea or flatulence.

According to the manufacturer's specifications, the recommended daily intake is 2 x 2 tablets (1.6 g polyglucosamine/day) for weight reduction. The number of lipid-adsorbent tablets (0.8 g polyglucosamine/day) ingested throughout the study is the recommended amount for weight maintenance.

The efficacy of polyglucosamine at different doses and also when administered concomitantly with a calorie-restricted diet was investigated in various studies (Sciutto and Colombo, 1995; Giustina and Ventura, 1995; Macchi, 1996; Colombo and Siutto, 1996; Veneroni et al., 1996; Schiller et al., 2001; Zahorska-Markiewicz et al., 2002; Mhurchu et al., 2004; Cornelli et al., 2008). The results showed that the additional administration of ß-1,4-poly-D-glucosamine within the scope of the common treatment for overweight and obesity (Caloric restriction) led to significantly more weight loss than with caloric restriction alone. In contrast, Pittler et al. (1999) reported that oral chitosan (1 g/day) did not reduce body weight in the absence of dietary alteration after 28 days of treatment.

In a Cochrane systematic review by Mhurchu et al. (2005), fourteen intervention studies including a total of 1131 participants were considered to assess the effects of chitosan on weight loss. Trials were only included if they were randomised controlled trials for a minimum of four weeks' duration. Study preparations containing chitosan achieved a significantly greater weight (-1.7 kg; $P < 0.001$), TC levels (-0.2 mmol/l; $P < 0.001$), SBP (-5.9 mmHg; $P < 0.001$) and DBP (-3.4 mmHg; $P < 0.001$) loss compared with placebo (Mhurchu et al., 2005). In previous clinical trials, the doses of polyglucosamine or chitosan administered were between 0.24 and 15.0 g per day (mean 3.7 g/day) and were, therefore, much higher than those provided during the intervention here (Macchi, 1996; Williams, 1998; Ho et al., 2001; Schiller et al., 2001; Zahorska-Markiewicz et al., 2002; Mhurchu et al., 2005; Mhurchu et al., 2004; Cornelli et al., 2008; Tapola et al., 2008; Anraku et al., 2009). In the study at hand, the administration of polyglucosamine (0.8 g/day) showed a further slight (-0.74 kg) but not significant weight loss compared with placebo. This is not consistent with the findings in previous studies in which higher dosages were given. However, it should be noted that only a few publications are available that indicate a dose-response relationship (Jaffer and Sampalis, 2007).

According to Jull et al. (2008), there is some evidence that chitosan is more effective than placebo in the short-term treatment of obesity. But it was widely criticized that many trials mentioned above have been limited by poor quality (Mhurchu et al., 2005; Jull et al., 2008). Overall results from high quality trials only demonstrated minimal effect from chitosan on body weight. In this study, weight reduction may be mainly due to the effect of the high-protein formula diet. This confirms established findings that meal replacements are a valid alternative dietary strategy in the treatment of obesity (Keogh and Clifton, 2005).

However, the add-on intake of polyglucosamine had a more favourable effect on the parameters of the glucose metabolism compared with placebo in this clinical trial. While the insulin levels were significantly lowered in the two groups, a significant decrease in HbA1c levels was detected in group F+LA but not in the placebo group. This verifiable effect of polyglucosamine should be seen as a protective effect with regard to diabetes risk and the risk of developing metabolic syndrome. Additionally, it should be mentioned that baseline HbA1c levels were within normal range (4.1-6.2% laboratory reference value) for both groups. It is worth considering whether there are any benefits of HbA1c reduction within the normal range. In patients with higher initial HbA1c values, each 1% decrease in mean HbA1c was associated with risk reduction for diabetic complications of 21% for any diabetes endpoint (Stratton et al., 2000).

The combined approach to weight loss (F+LA) demonstrated in particular its favourable effect on blood lipid levels. The present intervention study showed that the additional intake of polyglucosamine had a significant hypocholesterolemic effect. TC, LDL-C and TAG levels were significantly lowered.

The participants in group F+LA benefited from the intake of polyglucosamine more, especially with regard to the cardiovascular risk, compared with group F+P. This additional benefit is indeed highly relevant with respect to the metabolic syndrome and related disorders. An improvement of the lipid profile with the use of polyglucosamine had already been observed in previous clinical studies (Ylitalo et al., 2002; Bokura and Kobayashi, 2003; Mhurchu et al., 2005; Cornelli et al., 2008). Therefore, the intake of polyglucosamine led to a significant lowering of the TC and LDL-C levels, a reduction of TAG concentrations and to an increase of HDL-C levels. Compared with Mhurchu et al. (2005) the reduction of total cholesterol levels in this study was apparently higher (-0.39 mmol/l vs. -0.15 mmol/l) despite a lower dose.

According to the EFSA, the cause and effect relationship between the consumption of chitosan and the maintenance of normal blood LDL-cholesterol concentrations has been established. The EFSA Panel considers that in order to achieve the claimed effect, 3 g of chitosan should be consumed daily (EFSA, 2011b).

The study is limited in its findings in the following ways:

There was no estimation of the energy and nutrient intake at baseline and after twelve weeks. Although physical activity has been monitored throughout the study, the actual influence cannot be quantified. These factors could be act as confounders. The remaining powder of the formula diet was not weighed as there would have been great variability. To control compliance to the intake of the formula diet, concentration of 25-OH-D3 (Vitamin D) and folic acid were determined in the serum (data not shown). The nutrient supply for vitamin D and folic acid improved significantly in both groups with the intake of the formula. The changes in the two groups were not significantly different. Due to the number of dropouts, the sample size and possibly the effect size decreased. Probably more subjects should have been included at baseline.

This study may represent only a short-term effect over twelve weeks. With a view to lasting weight maintenance, long-term effects are required.

5. CONCLUSIONS

The clinical investigation demonstrates that the moderate application of a meal replacement strategy led to a significant loss of clinically relevant body weight within twelve weeks. The additional administration of lipid-adsorbent tablets containing polyglucosamine with a meal a day showed a further slight but not significant weight loss compared to placebo. More important than weight loss ought to be the fact, that this treatment method had favourable effects on carbohydrate and lipid metabolism and led to a significant reduction of HbA1c, insulin, TC, LDL-C, and TAG. Further examinations concerning a dose-response relationship are needed.

COMPETING INTERESTS

The authors declared no conflict of interest. The preparations used in this study were provided by Certmedica International GmbH, Aschaffenburg, Germany.

REFERENCES

Alberti, K.G., Eckel, R.H., Grundy, S.M., Zimmet, P.Z., Cleeman, J.I., Donato, K.A., Fruchart, J.C., James, W.P., Loria, C.M., Smith, S.C., Jr. (2009). Harmonizing the metabolic syndrome: a joint interim statement of the international diabetes federation task force on epidemiology and prevention; national heart, lung and blood institute; American Heart Association; World Heart Federation; International Atherosclerosis Society; and International Association for the Study of Obesity. Circulation 120, 1640-1645.

Anraku, M., Fujii, T., Furutani, N., Kadowaki, D., Maruyama, T., Otagiri, M., Gebicki, J.M., Tomida, H. (2009). Antioxidant effects of a dietary supplement: reduction of indices of oxidative stress in normal subjects by water-soluble chitosan. Food Chem Toxicol., 47(1), 104-109.

Apfelbacher, C.J., Cairns, J., Bruckner, T., Mohrenschlager, M., Behrendt, H., Ring, J., Kramer, U. (2008). Prevalence of overweight and obesity in East and West German children in the decade after reunification: population-based series of cross-sectional studies. J Epidemiol Community Health, 62, 125-130.

Ashley, J.M., St Jeor, S.T., Schrage, J.P., Perumean-Chaney, S.E., Gilbertson, M.C., McCall, N.L., Bovee, V. (2001a). Weight control in the physician's office. Arch. Intern. Med., 161, 1599-1604.

Ashley, J.M., St Jeor, S.T., Perumean-Chaney, S., Schrage, J., Bovee, V. (2001b). Meal replacements in weight intervention. Obes. Res., 9 Suppl 4, 312S-320S.

Bokura, H., Kobayashi, S. (2003). Chitosan decreases total cholesterol in women: a randomized, double-blind, placebo-controlled trial. Eur. J Clin. Nutr., 57, 721-725.

Brunner, E.J., Rees, K., Ward, K., Burke, M., Thorogood, M. (2007). Dietary advice for reducing cardiovascular risk. Cochrane Database Syst Rev., Issue 4, Art. No.: CD002128, doi: 10.1002/14651858.CD002128.pub3.

Carbajo, M.A., Castro, M.J., Kleinfinger, S., Gómez-Arenas, S., Ortiz-Solórzano, J., Wellman, R., García-lanza, C., Luque, E. (2010). Effects of a balanced energy and high protein formula diet (Vegestart complet®) vs. low-calorie regular diet in morbid obese patients prior to bariatric surgery (laparoscopic single anastomosis gastric bypass): a prospective, double-blind randomized study. Nutr Hosp., 25(6), 939-48.

Case, C.C., Jones, P.H., Nelson, K., O'Brian, S.E., Ballantyne, C.M. (2002). Impact of weight loss on the metabolic syndrome. Diabetes Obes. Metab., 4, 407-414.

Colombo, P., Siutto, A.M. (1996). Nutritional aspects of chitosan employment in hypocaloric diet. Acta Toxicol. Ther., XVII, 287-302.

Cornelli, U., Belcaro, G., Cesarone, M.R., Cornelli, M. (2008). Use of polyglucosamine and physical activity to reduce body weight and dyslipidemia in moderately overweight subjects. Minerva Cardioangiol., 56, 71-78.

Ditschuneit, H.H., Flechtner-Mors, M. (2001). Value of structured meals for weight management: risk factors and long-term weight maintenance. Obes. Res., 9 Suppl 4, 284S-289S.

Ditschuneit, H.H., Flechtner-Mors, M., Johnson, T.D., Adler, G. (1999). Metabolic and weight-loss effects of a long-term dietary intervention in obese patients. Am. J. Clin. Nutr., 69, 198-204.

EFSA. (2011a). Scientific Opinion on the substantiation of health claims related to very low calorie diets (VLCDs) and reduction in body weight (ID 1410), reduction in the sense of hunger (ID 1411), reduction in body fat mass while maintaining lean body mass (ID 1412), reduction of post-prandial glycaemic responses (ID 1414), and maintenance of normal blood lipid profile (1421) pursuant to Article 13(1) of Regulation (EC) No 1924/2006. EFSA Journal, 9(6), 2271.

EFSA. (2011b). Scientific Opinion on the substantiation of health claims related to chitosan and reduction in body weight (ID 679, 1499), maintenance of normal blood LDL-cholesterol concentrations (ID 4663), reduction of intestinal transit time (ID 4664) and reduction of inflammation (ID 1985) pursuant to Article 13(1) of Regulation (EC) No 1924/2006. EFSA Journal, 9(6), 2214.

Flechtner-Mors, M., Ditschuneit, H.H., Johnson, T.D., Suchard, M.A., Adler, G. (2000). Metabolic and weight loss effects of long-term dietary intervention in obese patients: four-year results. Obes. Res., 8, 399-402.

Gellner, R., Domschke, W. (2008). Epidemiology of obesity. Chirurg 79, 807-6, 818.

Giustina, A., Ventura, P. (1995). Weight-reducing regims in obese subjects: effects of a new dietary fiber integrator. Acta Toxicol. Ther., XVI, 199-214.

Hamdy, O., Zwiefelhofer, D. (2010). Weight management using a meal replacement strategy in type 2 diabetes. Curr. Diab. Rep., 10, 159-164.

Ho, S.C., Tai, E.S., Eng, P.H., Tan, C.E., Fok, A.C. (2001). In the absence of dietary surveillance, chitosan does not reduce plasma lipids or obesity in hypercholesterolaemic obese Asian subjects. Singapore Med J., 42, 6-10.

Jaffer, S., Sampalis, J.S. (2007). Efficacy and safety of chitosan HEP-40TM in the management of hypercholesterolemia: a randomized, multicenter, placebo-controlled trial. Altern Med Rev., 12, 265-273.

Jull, A.B., Ni, M.C., Bennett, D.A., Dunshea-Mooij, C.A., Rodgers, A. (2008). Chitosan for overweight or obesity. Cochrane Database Syst. Rev., CD003892.

Kaats, G.R., Michalek, J.E., Preuss, H.G. (2006). Evaluating efficacy of a chitosan product using a double-blinded, placebo-controlled protocol. J Am. Coll. Nutr., 25, 389-394.

Kanauchi, O., Deuchi, K., Imasato, Y., Shizukuishi, M., Kobayashi, E. (1995). Mechanism for the inhibition of fat digestion by chitosan and for the synergistic effect of ascorbate. Biosci. Biotechnol. Biochem., 59, 786-790.

Keogh, J.B., Clifton, P.M. (2005) The role of meal replacements in obesity treatment. Obesity Reviews, 6, 229-234.

Kloting, N., Stumvoll, M., Bluher, M. (2007). The biology of visceral fat. Internist (Berl)., 48, 126-133.

Macchi, G. (1996). A new approach to the treatment of obesity: chitosan's effects on body weight reduction and plasma cholesterol's level. Acta Toxicol. Ther., XVII, 303-320.

Mensink, G.B., Lampert, T., Bergmann, E. (2005). Overweight and obesity in Germany 1984-2003. Bundesgesundheitsblatt. Gesundheitsforschung. Gesundheitsschutz., 48, 1348-1356.

Mhurchu, C., Dunshea-Mooij, C.A., Bennett, D., Rodgers, A. (2005). Chitosan for overweight or obesity. Cochrane. Database Syst. Rev., CD003892.

Mhurchu, C.N., Poppitt, S.D., McGill, A.T., Leahy, F.E., Bennett, D.A., Lin, R.B., Ormrod, D., Ward, L., Strik, C., Rodgers, A. (2004). The effect of the dietary supplement, Chitosan, on body weight: a randomised controlled trial in 250 overweight and obese adults. Int. J Obes. Relat Metab Disord., 28, 1149-1156.

Moura, E.C., Claro, R.M. (2011). Estimates of obesity trends in Brazil, 2006-2009. Int. J Public Health., DOI 10.1007/s00038-011-0262-8

Noakes, M., Foster, P.R., Keogh, J.B., Clifton, P.M. (2004). Meal replacements are as effective as structured weight-loss diets for treating obesity in adults with features of metabolic syndrome. J Nutr., 134, 1894-1899.

Pittler, M.H., Abbot, N.C., Harkness, E.F., Ernst, E. (1999). Randomized, double-blind trial of chitosan for body weight reduction. Eur. J Clin. Nutr., 53, 379-381.

Redmon, J.B., Reck, K.P., Raatz, S.K., Swanson, J.E., Kwong, C.A., Ji, H., Thomas, W., Bantle, J.P. (2005). Two-year outcome of a combination of weight loss therapies for type 2 diabetes. Diabetes Care, 28, 1311-1315.

Rodriguez, M.S., Albertengo, L.E. (2005). Interaction between chitosan and oil under stomach and duodenal digestive chemical conditions. Biosci. Biotechnol. Biochem., 69, 2057-2062.

Rolland, C., Hession, M., Murray, S., Wise, A., Broom, I. (2009). Randomized clinical trial of standard dietary treatment versus a low-carbohydrate/high-protein diet or the LighterLife Programme in the management of obesity. J Diabetes, 1(3), 207-217.

Rothacker, D.Q., Staniszewski, B.A., Ellis, P.K. (2001). Liquid meal replacement vs traditional food: a potential model for women who cannot maintain eating habit change. J Am. Diet. Assoc., 101, 345-347.

Schiller, R.N., Barrager, E., Schauss, A.G., Nichols, E.J. (2001). A randomized, double-blind, F+P-controlled study examining the effects of a rapidly soluble chitosan dietary supplement on weight loss and body composition in overweight and mildly obese individuals. Journal of the American Nutraceutical Association, 4, 42-49.

Sciutto, A.M., Colombo, P. (1995). Lipid-lowering effect of chitosan dietary integrator and hypocaloric diet in obese subjects. Acta Toxicol. Ther., XVI, 215-230.

Shaw, K., Gennat, H., O'Rourke, P., Del Mar, C. (2006). Exercise for overweight or obesity. Cochrane Database Syst Rev., Issue 4. Art. No.: CD003817. doi: 10.1002/14651858.CD003817.pub3.

Smith, T.J., Sigrist, L.D., Bathalon, G.P., McGraw, S., Karl, J.P., Young, A.J. (2010). Efficacy of a meal-replacement program for promoting blood lipid changes and weight and body fat loss in US Army soldiers. J Am. Diet. Assoc., 110, 268-273.

Stratton, I.M., Adler, A.I., Neil, H.A., Matthews, D.R., Manley, S.E., Cull, C.A., Hadden, D., Turner, R.C., Holman, R.R. (2000). Association of glycaemia with macrovascular and microvascular complications of type 2 diabetes (UKPDS 35): prospective observational study. BMJ, 321(7258), 405-12.

Tapola, N.S., Lyyra, M.L., Kolehmainen, R.M., Sarkkinen, E.S., Schauss, A.G. (2008). Safety aspects and cholesterol-lowering efficacy of chitosan tablets. J Am Coll Nutr., 27, 22-30.

Veneroni, G., Veneroni, F., Contos, S., Tripodi, S., De Bernardi, M., Guardino, C., Marletta, M. (1996). Effect of a new chitosan dietary integrator and hypocaloric diet on hyperlipidemia and overweight in obese patients. Acta Toxicol. Ther., XVII, 53-70.

Williams, A.R. (1998). A double-blind, placebo-controlled evaluation of the effects of RW94 on the body weight of both overweight and obese healthy volunteers. Curr Med Res Opin., 14, 243-249.

Yao, H.T., Chiang, M.T. (2006). Effect of chitosan on plasma lipids, hepatic lipids, and fecal bile acid in hamsters. Journal of Food and Drug Analysis, 14, 183-189.

Yip, I., Go, V.L., DeShields, S., Saltsman, P., Bellman, M., Thames, G., Murray, S., Wang, H.J., Elashoff, R., Heber, D. (2001). Liquid meal replacements and glycemic control in obese type 2 diabetes patients. Obes. Res., 9 Suppl 4, 341S-347S.

Ylitalo, R., Lehtinen, S., Wuolijoki, E., Ylitalo, P., Lehtimaki, T. (2002). Cholesterol-lowering properties and safety of chitosan. Arzneimittelforschung., 52, 1-7.

Zahorska-Markiewicz, B., Krotkiewski, M., Olszanecka-Glinianowicz, M., Zurakowski, A. (2002). Effect of chitosan in complex management of obesity. Pol. Merkur Lekarski,

5

Valorisation of Apple Peels

Laura Massini[1*], Daniel Rico[2], Ana Belen Martín Diana[2] and Catherine Barry-Ryan[1]

[1]School of Food Science and Environmental Health, Dublin Institute of Technology, Cathal Brugha Street, D1, Dublin, Ireland.
[2]Agro Technological Institute of Castilla y Leon (ITACYL), Government of Castilla y Leon, Finca Zamadueñas, Valladolid, Spain.

Authors' contributions

The experimental study was designed in collaboration between all authors. Author LM managed the literature searches, the analyses of the study and wrote the first draft of the manuscript. All authors read and approved the final manuscript.

ABSTRACT

The peels of processed apples can be recovered for further food applications. Limited information on the valorisation of this type of waste is available for cooking varieties, e.g. cv Bramley's Seedling. Extracts from fresh or dried (oven-dried or freeze-dried) peels were obtained with solvents of different polarity (aqueous acetone or ethanol) and assayed for their total phenolic content and antioxidant capacity; their antiradical power was compared to herb extracts. The dried peels were also characterised as bulk powders by assessing their nutritional value and total phenolic content. High amounts of ascorbic acid (up to 4 mg/g, dry weight) and polyphenols (up to 27 mg gallic acid equivalents/g, dry weight) were found in the peels, with the latter contributing significantly to the antioxidant capacity; the nutrient profile was low in protein (less than 10%, w/w) and total dietary fibre content (less than 40%, w/w). Higher yields of phenolic antioxidants were recovered with acetone from freeze-dried peels; the resulting extracts had equivalent antioxidant power to oregano leaves (*Origanum vulgare* L.). The combination of oven-drying/ethanol led to lower recovery yields of phenolic antioxidants; however, these conditions could increase the feasibility of the extraction process, leading to antioxidant extracts with lower energy or cost input, and higher suitability for further food use.
The recovery of phenolic antioxidants from the peels of processed apples could be a

Corresponding author: Email: laura.massini@dit.ie

valuable alternative to traditional disposal routes (including landfill), in particular for cooking varieties.
The recycling process could enhance the growth of traditional culinary apple markets in UK and Ireland thanks to the new business opportunities for the peel-derived materials.

Keywords: Waste valorisation; cooking apples; peel polyphenols; antioxidant value.

1. INTRODUCTION

There is an increasing interest about natural plant extracts (i.e. botanicals) in novel food applications, as nutraceutical ingredients [1] or natural preservatives [2] and antioxidants [3,4,5]. Various agri-food waste and by-products have been screened for the recovery of natural phenolic antioxidants [6]. The recovery of valuable materials is a strategy of waste minimisation [7]. Some nutraceutical products have been developed from grape waste or apple peels, and marketed for the functional markets of Japan and USA [8,9]. In Europe, the use of botanicals such as vegetable and fruits, herbs and spices, herbal teas and infusions, and herbs is allowed in food and beverages for taste or functional purposes (e.g. guarana, gentian, etc.) [1]; however, the functional applications of many botanicals have not yet received the scientific opinions of the European Food Safety Authority (EFSA) [10].

Apples are important dietary sources of phenolic compounds and have strong antioxidant capacity compared to other fruits [11]. Apple polyphenols have various *in vitro* bioactivities, possibly in combination with dietary fibre (i.e. reduced risk of coronary heart disease) [12]. Higher amounts of polyphenols, in particular flavonol glycosides, are generally found in the skin of the fruit, compared to the pulp [13].

Some studies have reported about the recycling of apple peels as a source of phenolic compounds and/or dietary fibre; depending on the compounds, different peel waste-derived materials were developed (Table 1). The apple peels were preferably processed into a dried and pulverised bulk material for fibre formulation or nutraceutical use. Phenolics were extracted with organic solvents (or aqueous mixtures thereof) and then characterised for their potential health benefits. The second recycling option involved the preparation of crude or purified mixtures of phenolic antioxidants and/or their formulation in nutraceutical or functional food applications. To the best of our knowledge, the preparation and characterisation of apple peel extracts for food stabilisation or preservation has not been studied.

In the preparation and characterisation of plant waste-derived materials, conditions such as the drying and the liquid extraction of phenolic compounds have an impact onto the feasibility of the recycling process (i.e. energy consumption and cost input), and further applications of the recovered ingredient [14]. For example, the extracts from apple peels developed in [15] were obtained with methanol; therefore they could not be tested in food systems. Ethanol and water should be preferred over methanol in view of food applications [16]. Freeze-drying, which is advantageous for heat sensitive materials, also requires higher energy consumption and initial and maintenance costs than oven-drying or air-drying, therefore its use could be limited in the industry [17].

Table 1. Recycling of apple peel-derived materials: processing conditions (drying; extraction solvent); target compounds; and further applications

Peel-derived materials	Preservation conditions		Extraction solvent (phenolic compounds)	Applications	Target compounds	References
	Pre-drying treatments (peel material)	Drying				
Bulk peel powders	N/A	Drum-drying;	70% Acetone (v/v)	Fibre formulation/ Functional foods	Dietary fibre and phenolic compounds	[18]
	Water blanching;	Oven-drying (60°C, with air circulation)	Methanol	Fibre formulation/ Functional foods	Dietary fibre and phenolic compounds	[19]
	Water blanching; ascorbic acid dip	Freeze-drying; air-drying; oven-drying (at 40/60/80°C, no air circulation)	80% Acetone or 80% ethanol (v/v)	Nutraceuticals	Phenolic compounds	[20]
	N/A	Freeze-drying	Methanol	Functional foods	Phenolic compounds	[15]
Antioxidant peel extracts	N/A	N/A	N/A	Functional foods	Phenolic compounds	[21]a
	N/A	N/A	Ethanol or methanol	Nutraceuticals	Phenolic compounds	[22]
	N/A	Freeze-drying	80% Acetone (v/v)	Nutraceuticals	Phenolic compounds	[23]

a In this study, the apple peel extract was commercially available; the conditions used for its preparation were not described. N/A: not applicable.

The diversion of the peel waste from traditional disposal routes (landfertilising, feedstock, or landfill) towards more valuable food applications could favour the sustainable development of the culinary apple markets in the British Isles that are primarily based on cv Bramley's Seedling. This variety is known for the sole purpose of cooking, i.e. processed into sauce or puree, or used for home baking. Due to changes in the lifestyle, at the end of the 90's the fresh sector has narrowed in UK [24]; the same trend has occurred in Ireland, with the consequent overproduction at low farm gate prices [25]. In the absence of official statistics about the waste generated, it was estimated that 300 tonnes of peels could be discarded annually by processing lines in Ireland [26], assuming a yield of 11% (w/w) of peels from the whole apple. Another 5,000 tonnes of peels could be generated from the amount of processed lines in UK.[1]

The peels and/or pulp of cooking apples were assessed for their phenolic content in order to establish their dietary significance [27,28]. However, few studies have investigated their recovery for valuable applications. Polyphenols were extracted from the pomace as potential nutraceutical compounds [29]. The contribution of the skin to the extractable phenolics from the pomace was studied in comparison to the peeled fruit, distinguishing among soluble and insoluble bound components in view of further applications [30].

In the present study, different approaches for the preparation of peel-derived materials (bulk powders or extracts) with nutritional and/or antioxidant value from cv Bramley's Seedling apple (origin: Ireland) were investigated with the aim of establishing an optimal recovery process for further food use. The recycling value of these materials was compared to other plant-based products already developed for food applications (i.e. from the peels of different apple varieties; or herb leaves). Processing conditions (drying and/or extraction solvent) with different energetic or cost input were compared with the aim of defining a feasible recycling process with increased industrial applications. This valorisation approach could be applied to other processed apples in order to increase the type of waste-derived products recovered from solid fruit waste.

2. MATERIALS AND METHODS

2.1 Chemicals

Chemicals were purchased from Sigma-Aldrich (Ireland) and included: sodium nitrite; sodium carbonate; ferric chloride; aluminium chloride hexahydrate; 2.0 N Folin-Ciocalteu's phenol reagent; 2,4,6-tri(2-pyridyl)-s-triazine (TPTZ); 2,2-diphenyl-1-picrylhydrazyl (DPPH); Celite, acid-washed; enzymes for the digestion of the dietary fibre: amyloglucosidase from *Aspergillus niger*; protease from *Bacillus licheniformis*; α-amylase (heat stable) from *Bacillus licheniformis*; and the standards: (+)-catechin hydrate; gallic acid and L-ascorbic acid.

2.2 Plant Material

Two batches of apples (i.e. 3-5 kg per batch) (*Malus domestica* Borkh. cv. Bramley's Seedling) were purchased from a local store (Dublin, Ireland) between October 2007 and April 2008. According to the information provided by the retailer, the apples were grown in Co. Armagh, Northern Ireland, harvested in late August/September and made available throughout the year thanks to storage facilities (under controlled atmosphere).

---

[1] http://www.bramleyapples.co.uk

The purchased apples were stored at 4°C in a polyethylene film, until processing. The apples were washed under tap water, dried by patting on a paper cloth and weighed. The peels were manually removed with a hand-peeler. Five grams of fresh peels were collected in triplicate from each batch of apples and immediately assayed. The remaining peels were oven-dried at 60 ± 2 °C (OD) on stainless steel trays in a ventilated oven (BS Oven 250, Weiss Gallenkamp, UK) or freeze-dried (FD) in a Micro Modulyo E-C Apparatus (Davidson & Hardy, USA) until a constant weight was achieved, in the dark. After drying, the samples were pulverised in a coffee grinder and the resulting powders were stored in amber bottles at -20°C until analysis.

2.3 Experimental Design

The experimental design included the preparation of peel extracts from oven-dried samples with 80% ethanol, or freeze-dried peels with 80% acetone. The drying and solvent systems were studied under these combinations (i.e. freeze-drying/acetone; and oven-drying/ethanol) with the purpose of comparing conditions with less or more favourable impact onto the feasibility of the recovery process. The resulting extracts were compared to fresh samples extracted under similar conditions in order to assess the effect of processing onto the phenolic content and antioxidant capacity of the peels. Oregano and rosemary leaf extracts were prepared from herbs purchased from a local store and used as reference plant extracts with established food applications [2]. The dried and pulverised peels were also characterised as bulk materials (i.e. nutritional value and total phenolic content). Soluble phenolic compounds were extracted with acetone or ethanol from dried peels (oven-dried or freeze-dried) and further quantified. The colour and free acidity of the powders were assessed because of their potential sensorial impact in further food formulation.

2.4 Characterisation of Bulk Peel Powders

2.4.1 Proximate analysis

The proximate analysis was carried out according to official methods [31]: moisture content (Method 930.04); ash content (Method 930.05); protein content (Method 920.152); fat content (Method 983.23, with petroleum ether); ascorbic acid content (Method 967.21). The total dietary fibre (TDF) was determined according to [32]. Sugars were extracted from the plant matrix using 80% ethanol (v/v) under boiling conditions and quantified as glucose equivalents (g/100 g) using the phenol-sulphur method [33]. The analyses were done in triplicate and expressed on a dry weight basis (DW).

2.4.2 Free titratable acidity

For the free titratable acidity, 1 g of peel powder was boiled for 10 mins in 20 mL of distilled water and filtered through a Büchner funnel. The free titratable acidity was measured according to [31] (Method 942.15.b).

2.4.3 Colour

The CIELAB* colour (L*; a*; b* values) of the powders was measured in triplicate using ColorQuest®Xe (HunterLab, USA) applying the reflectance method: 10° observer; D65 illuminant. The instrument was calibrated with standard white and black tiles. The colour values were expressed as: L* = lightness (from 0 to 100); a* = redness/greenness (from +a* to –a*); b* = yellowness/blueness (from +b* to –b*).

2.5 Characterisation of Peel Extracts

2.5.1 Extraction of phenolic compounds

Crude mixtures of soluble polyphenols were obtained in triplicate from fresh or dried peels, using a procedure previously described with minor modifications [20]. For the dried peels, ~1 gram of powder was homogenised (Ultra-Turrax T25, IKA Laborteck, Germany) with 40 g of chilled aqueous 80% ethanol or 80% acetone (v/v) at 9500-13500 min^{-1} for 5 min. The obtained slurry was filtered under vacuum. The remaining solids were added to 15 mL solvent and extracted again, homogenising for 1 min. For the fresh peels, 5 g of sample was blended in a portable mini blender (dj2000 Illico Mini Chopper, Moulinex, France) with 40 g of solvent for 3 min, and then filtered through N.6 Whatman paper in a Büchner funnel. In the last filtration step, for both fresh and dried samples, another 15 mL of solvent was poured onto the filter cake. During the extraction, the extracts were kept chilled in an ice bath, in the dark. Homogenisation was stopped after one minute, waiting at least another minute before resuming. The filtrates were collected and the organic solvent was removed at 40°C using a Büchi rotavapor, until the aqueous phase remained. The concentrated extracts were brought to the volume of 25 mL with distilled water, filtered through N.1 Whatman paper, and stored at -20°C in the dark. Before analysis, they were thawed, centrifuged at 8,000 rpm for 15 min, filtered through 0.45 μm PTFE (Acrodisc, Pall, UK) membrane disc filter, and brought up to the volume of 50 mL with distilled water.

2.5.2 Total phenolic content

The total phenolic content (TPC) was assessed using Folin-Ciocalteu assay [34]. Volumes of 0.5 mL of distilled water and 0.125 mL of sample were added to a test tube. A volume of 0.125 mL of 2.0 N Folin-Ciocalteu reagent was added and allowed to react for 6 min. Then, 1.25 mL of a 7% sodium carbonate solution (v/v) was added to the mixture and allowed to stand for 90 min in the dark, for colour development. Before reading the absorbance at 760 nm in a spectrophotometer (Spectronic 1201, Milton Roy, USA), the mixture was diluted up to 3 mL with distilled water. Gallic acid solutions were used for the standard calibration curve and the total phenolic content was expressed as mg gallic acid equivalents (GAE)/g or 100 g peels (dry weight or fresh weight basis, DW or FW). All measurements were carried out in triplicate.

2.5.3 Total flavonoid content

The total flavonoid content (TFC) was assessed using aluminium-chloride assay [35]. A volume of 0.25 mL of sample was added to a test tube containing 1.25 mL of distilled water. An aliquot of 0.075 mL of 5% sodium nitrite solution (w/v) was added to the mixture and allowed to stand for 5 min. Then, the addition of 0.15 mL of 10% aluminium chloride (w/v) developed a yellow flavonoid-aluminium complex. After 6 min, 0.5 mL of 4.3% NaOH (w/v) was added. The absorbance was measured immediately in a spectrophotometer (Spectronic 1201, Milton Roy, USA) at 510 nm and compared to a standard curve of (+)-catechin solutions. The flavonoid content was expressed as mg catechin equivalents (CE)/g peels (FW). All measurements were carried out in triplicate.

2.5.4 Ferric reducing antioxidant power

The antioxidant capacity was evaluated using a modified FRAP assay procedure based on a previously published protocol [36]. A freshly prepared FRAP-reagent (25 mL acetate buffer, 300 mM, pH 3.6 + 2.5 mL 10 mM TPTZ (2,4,6-tripyridyl-5-triazine) in 40 mM HCl + 2.5 mL 20 mM $FeCl_3 \cdot 6$ H_2O) was heated in water bath at 37°C for 5 min before being transferred (0.9 mL) into tubes containing 0.1 mL of plant extracts. The tubes were left in water bath at 37°C for 40 minutes. The absorbance was then measured at 593 nm in a spectrophotometer (Spectronic 1201, Milton Roy, USA). The antioxidant capacity was compared to standard L-ascorbic acid through a calibration curve, and expressed as mg ascorbic acid equivalents (AAE)/g peels (FW), which was also referred to as AEAC (ascorbic acid equivalent antioxidant capacity). All measurements were carried out in triplicate.

2.5.5 Radical scavenging capacity

The radical scavenging capacity was measured against the synthetic radical compound $DPPH^{\bullet}$ [37]. A volume of 0.1 mL of diluted extracts (bulk; 1:2; 1:5; 1:10; 1:20; 1:50) was added in a reaction vessel containing 0.9 mL of a freshly prepared $DPPH^{\bullet}$ solution (0.08 mM in 96% ethanol, v/v); the reaction was allowed to run for at least 30 minutes. The decrease in absorbance of the samples was read at 515 nm against a blank of distilled water in a spectrophotometer (Spectronic 1201, Milton Roy, USA) and compared to that of a control solution of $DPPH^{\bullet}$ prepared with 0.1 mL of distilled water.

The % Reduced $DPPH^{\bullet}$ was calculated using the following equation:

$$\% \text{ Reduced } DPPH^{\bullet} = [(1 - \text{Abs sample})/\text{Abs control}] * 100$$

The % Reduced values were expressed as AEAC (mg AAE/g peels, FW) by comparison with a standard calibration curve with ascorbic acid. The IC_{50} value (i.e. concentration of plant extract that reduces by 50% the initial concentration of the radical form of $DPPH^{\bullet}$ in the reaction mixture) was calculated from the curves of sample concentration (as mg/mL, FW) vs. % Reduced $DPPH^{\bullet}$. The values were expressed as Antiradical Power (ARP) = $1/IC_{50}$ (mL/g sample, FW) according to [38]. For the preparation of plant extracts with reference antiradical power, fresh leaves of oregano (OR) and rosemary (ROS) were purchased from a local store (Dublin, Ireland) and oven-dried at 60°C ± 2°C in a ventilated air oven (Weiss Gallenkamp BS Oven 250, UK) until constant weight was achieved, in the dark. The samples were pulverised using a mortar and a pestle. Rosemary (5 g) and oregano (2 g) leaf powders were extracted with 95% ethanol (v/v) homogenising for 2 minutes [39]. The resulting ROS and OR extracts were filtered through N°6 Whatman filter paper using a Büchner funnel, under vacuum. The filtrates were collected and further evaporated in a rotary evaporator at 40°C under vacuum, until 20% of the original volume remained. The extracts were stored in amber glass bottles at -20°C until analysis.

2.6 Statistical Analysis

Statistical analysis was conducted using StatGraphics Centurion XV (Statpoint Technologies Inc., USA) and GraphPad v. 5.01 for Windows (GraphPad Software Inc., USA). Normal data was tested for significance using the one-way ANOVA (*LSD* post-hoc test), and *F*-test as appropriate. A regression analysis was also carried out. For all the statistical tests, the significance level taken was $P<0.05$.

3. RESULTS AND DISCUSSION

3.1 Bulk Peel Powders

The characteristics of the powders obtained under different drying conditions were studied and further compared (Table 2). Regardless of the drying method, the powders generally had reduced protein content (less than 5%), making them a poor animal feed. They had high content of total carbohydrates (up to 80%, w/w). When compared to peel materials already developed from dessert varieties, e.g. cv Granny Smith [18], cv Northern Spy or cv Ida Red [19], the powders from Bramley apple peels had lower total dietary fibre (less than 40%, w/w, DW). They also had high acidity (almost 4-fold higher than in the peels of cv Granny Smith), which could negatively impact the sensorial characteristics in further food formulations. The ascorbic acid content was high, with values ranging from 3.0 to 4.4 (mg/g, DW); values between 0.7–3.4 mg/g were reported in the peels of various dessert apples [40].

Table 2. Physical and chemical characteristics of bulk peel powders as affected by the drying method

Parameter (%, w/w)	Drying method	
	OD	FD
Total ash	$2.23^a \pm 0.10$	$2.49^a \pm 0.44$
Total fat	$3.83^b \pm 0.23$	$6.61^a \pm 0.82$
Total protein	$5.07^a \pm 0.32$	$5.36^a \pm 0.19$
Total dietary fibre	$35.38^a \pm 2.22$	$32.49^a \pm 0.10$
Total sugars (as glucose)	$46.00^a \pm 8.27$	$40.36^a \pm 3.03$
Free titratable acidity (% malic acid, w/v)	$8.52^a \pm 0.11$	$8.16^a \pm 0.76$
Ascorbic acid (mg/g)	$3.01^b \pm 0.30$	$4.42^a \pm 0.20$
Colour		
L*	$71.3^b \pm 0.6$	$74.3^a \pm 0.2$
a*	$1.9^a \pm 0.2$	$-6.6^b \pm 0.1$
b*	$30.5^b \pm 0.3$	$34.6^a \pm 0.1$

Values were expressed as mean ± SD (n = 6) on a dry weight basis, considering an average residual moisture content of 7.5% and 9.0% for oven-dried (OD) and freeze-dried (FD) peels, respectively. Different superscript letters in each row denoted significant difference (P<0.05) between samples.

Some physical and chemical parameters were significantly affected by the drying system (Table 2). In particular, the thermal drying (e.g. oven-drying) produced a significant reduction of the fat and ascorbic acid content of the powders in comparison to freeze-drying. The oven-dried powders poorly retained the colour of the fresh peels in comparison to freeze-dried samples, and their colour had significant (P<0.05) lower greenness and yellowness values.

The drying system also influenced significantly (P<0.001) the yield of total phenolic compounds (calculated as TPC) in the final powders (Table 3). The yield also depended on the organic solvent used for their extraction (P<0.001). The thermal decomposition of the lipid substances in the skin could be associated to an increased oxidative damage of its natural antioxidants.

Table 3. Total phenolic content of oven-dried and freeze-dried bulk peel powders (extracted with different organic solvents)

Drying system	Extraction solvent	Total phenolic content (mg GAE/g, DW)
Freeze-drying (FD)	Acetone (Ac)	27.04 ± 1.76
	Ethanol (Et)	21.93 ± 0.36
Oven-drying (OD)	Acetone (Ac)	21.75 ± 0.36
	Ethanol (Et)	17.97 ± 0.42
Main effects	F-test LSD$_{0.05}$ = 1.24	Mean
Drying system	***	24.97 (FD) 20.04 (OD)
Extraction solvent	***	24.78 (Ac) 20.23 (Et)

*** indicated a highly significant effect (P<0.001). TPC values were expressed as mean ± SD (n = 6). GAE: gallic acid equivalents.

The loss of phenolic compounds during oven-drying was reported for various plants [6]. Natural antioxidants are normally accumulated in the skin in order to supply their antioxidant protection [40]. Phenolics could be regenerated by non-enzymatic reactions with ascorbate in the apple fruit [41]. The TPC values of the Bramley apple peels were in agreement with results already reported for this variety [27].

3.2 Peel Extracts

3.2.1 Phenolic yield

The total phenolic (TPC) and flavonoid (TFC) contents of fresh and dried peels extracted with different solvents were compared (Table 4). With regard to the same solvent, dried peels had similar TPC than fresh samples, but their TFC was significantly different (P<0.05).

Table 4. Phenolic content and antioxidant capacity of fresh and dried peels extracted with the same type of solvent

Parameter (mg/g peels, FW)	Extraction solvent	Peels	
		Fresh	Dried[1]
TPC (as GAE)	Acetone	7.68[a] ± 0.74	7.63[a] ± 0.17
	Ethanol	6.35[b] ± 0.76	5.86[b] ± 0.35
TFC (as CE)	Acetone	5.34[a] ± 0.48	4.51[b] ± 0.10
	Ethanol	4.76[b] ± 0.47	4.03[c] ± 0.06
FRAP (as AEAC)	Acetone	13.26[a] ± 0.88	13.92[a] ± 0.29
	Ethanol	9.88[b] ± 1.66	10.43[b] ± 1.34
Radical scavenging capacity (DPPH) (as AEAC)	Acetone	12.11[a] ± 1.22	10.43[b] ± 1.34
	Ethanol	9.15[c] ± 0.61	7.27[d] ± 0.64

[1] Freeze-dried (extracted with acetone); oven-dried (extracted with ethanol).
Values were expressed as mean ± SD (n=6). Different superscript letters indicated significant difference (P<0.05) between fresh and dried samples extracted with the same type of solvent (within row). TPC: total phenolic content, expressed as gallic acid equivalents (GAE); TFC: total flavonoid content, expressed as catechin equivalents (CE); FRAP: ferric reducing antioxidant power, expressed as ascorbic acid equivalents (AEAC); Radical scavenging capacity against DPPH, expressed as ascorbic acid equivalents (AEAC).

These findings suggested that some flavonoids were lost during the processing of the peels, while other phenolics (i.e. conjugated) could be released after hydrolysis of the cell wall linkages, thus contributing to the yield of total phenolics. Most of the conjugated phenolics in apples are esters of hydroxycinnammic acids [42].

With regard to the extraction solvent, acetone extracted higher amounts of phenolic compounds than ethanol. In particular, the yield of phenolic compounds with ethanol was nearly 20% less than with acetone. The solubility of plant phenolics in solvents such as ethanol or water is due to glycosilated forms than are more water-soluble than the related aglycones. A solvent of lower polarity, such as acetone, can favour the extraction of flavonoids of low-medium polarity (procyanidins) that remain otherwise bound to the alcohol-insoluble matrix in apples [43].

3.2.2 Antioxidant capacity

The ascorbic acid equivalent antioxidant capacities (AEAC) of the processed samples were compared to those of fresh samples extracted under the same solvent conditions (Table 4). The radical scavenging capacity (for DPPH$^•$) reduced significantly ($P<0.05$) after the processing of the peels, while the ferric reducing antioxidant power was not affected. These findings suggested that the redox potential (FRAP) of the fresh sample was maintained during processing because the amount of total reducing substances (including total polyphenols, TPC) remained stable possibly as a result of released hydroxycinnamic acids otherwise bound in the fresh tissue [20]. On the contrary, the radical scavenging capacity of the processed mixture lowered in comparison to fresh samples, possibly in response to the loss of flavonoid compounds (TFC). In particular, it is believed that the loss of oligomeric procyanidins, i.e. indicated as the most powerful antioxidants in apples [44], could influence significantly the radical scavenging capacity of the processed samples, as it is known that the number and substitution patterns of hydroxyl groups on the flavonoid structure is crucial for their radical scavenging capacity [45]. The two antioxidant assays, FRAP and DPPH, could respond differently to the antioxidant mixtures as they are based on different antioxidant mechanisms [46,47]. With regard to the solvent, the extracts obtained with acetone showed significantly higher antioxidant capacity ($P<0.05$) than those obtained with ethanol. This was explained as due to the solubilisation of higher amounts of phenolic compounds (especially flavonoids). The FRAP capacities of fresh and dried peels from cv. Bramley's Seedling were in agreement with data reported for dessert apples [13]. To the best of our knowledge, no AEAC values measured by the DPPH assay have been reported in literature for other apple peels.

3.2.3 Antiradical power

The Antiradical Power (ARP) of apple peel extracts was compared to oregano and rosemary leaf extracts (Fig. 1).

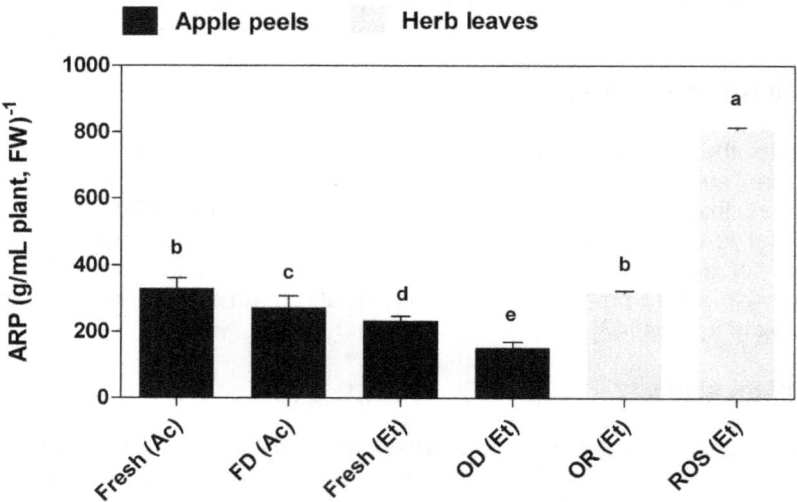

Fig. 1. Antiradical power of apple peel and herb leaf extracts.
Different superscript letters denoted significant difference (P<0.05) among samples. Drying: oven-drying (OD); freeze-drying (FD). Extraction solvent: acetone (Ac); ethanol (Et). Herbs: oregano (OR); rosemary (ROS).

The peel extracts obtained with acetone had similar antioxidant capacity than oregano leaf extracts. Rosemary extract had the strongest ARP ($P<0.05$) amongst the plant extracts investigated. Fresh peels had IC_{50} values of 4.28 ± 0.23 and 3.04 ± 0.27 mg peels/mL (FW) when extracted with ethanol and acetone, respectively. Dried peels had IC_{50} values of 6.51 ± 0.84 and 3.72 ± 0.48 mg peels/mL (FW), when extracted with ethanol and acetone, respectively. Kondo et al. [48] reported for the skin of dessert and cider apples IC_{50} values lower than 5 mg peels/mL (in the reaction mixture, FW), that is ARP values higher than 200 mL/g. The ARP values for fresh peels of cv. Bramley's Seedling in this study were 234 ± 13 and 331 ± 30 mL/g peels (in the reaction mixture, FW), for the extracts obtained with ethanol and acetone, respectively.

Oregano and rosemary leaf extracts had IC_{50} values of 3.13 ± 0.04 and 1.89 ± 1.12 mg herb/mL (FW); these values were equivalent to 0.39 and 0.16 mg herb/mL on DW basis, assuming an average moisture content of 86%, w/w, which were consistent with previous data reported in literature [49].

3.2.4 Regression analysis between antioxidant capacity and phenolic content

A regression analysis between the antioxidant capacity and the phenolic content of the peels was carried out (Table 5). The Pearson correlation coefficients were strongly significant ($P<0.01$) between the variables. However, it was observed a higher deviation from linearity in the regression values (*r-square*<0.66) of the whole peels (fresh + dried, n = 18) compared to dried samples (n = 12). This could indicate that reducing substances other than polyphenols (e.g. ascorbic acid) were extracted from fresh samples and contributed to the antioxidant capacity together with phenolics. In agreement with this hypothesis, the relationship between AEAC (measured as FRAP) and the total flavonoid content (*r-square*<0.34) was weak; while the radical scavenging capacity was better correlated with the total flavonoid content (*r-square*>0.63).

Table 5. Regression analysis between antioxidant capacity and phenolic content of apple peels

Antioxidant capacity (as AEAC)	Total phenolic content		Total flavonoid content	
Fresh+Dried	Corr.	r-square	Corr.	r-square
FRAP	**	(0.66)	**	(0.34)
DPPH	**	(0.47)	**	(0.63)
Dried	Corr.	r-square	Corr.	r-square
FRAP	**	(0.76)	**	(0.48)
DPPH	**	(0.63)	**	(0.69)

** indicated a very significant correlation between the variables (P<0.01); the linear regression fit for the correlated data was reported in brackets (R-square). AEAC: ascorbic acid equivalent antioxidant capacity; Corr.: Pearson's correlation.

In the dried samples, the contribution of phenolic compounds to the antioxidant capacity increased above 70%, particularly for flavonoids and their radical scavenging capacity, thus indicating the possible reduction of co-extracted substances, such as ascorbic acid. Results previously reported [27] for Bramley apple indicated a weak linear correlation between the antioxidant capacity (as FRAP) and the total phenolic content (*r-square*<0.58).

4. CONCLUSIONS

- The recycling value of the peels from cv. Bramley's Seedling depended on its high levels of natural antioxidants, in particular phenolic compounds that contributed significantly to its antioxidant capacity.
- The recovery of target phenolic antioxidants (especially flavonoids) could be lowered by the processing, i.e. cutting; drying and pulverising; however, during the processing, phenolic compounds conjugated in the fresh plant matrix could be released with a consequent increase of the redox potential and total phenolic content of the resulting extracts.
- The drying system and the organic solvent used for the phenolic recovery affected their extraction yield, consequently their antioxidant capacity. Freeze-drying protected the antioxidant value better than oven-drying, while acetone favoured the solubilisation of higher amounts of phenolic compounds than ethanol. The resulting extracts had equivalent antioxidant power to oregano leaf extract.
- The use of oven-drying/ethanol for the phenolic recovery could lead to extracts with lower antioxidant value compared to freeze-drying/acetone but with enhanced food applications.
- Further investigation on the isolation of antioxidant phenolic compounds from the peels of Bramley's Seedling apple for future food applications is desirable.

ACKNOWLEDGMENTS

The authors would like to acknowledge the financial support of the DIT Strand III 2007-2010 for the carrying out of this project.

COMPETING INTERESTS

Authors have declared that no competing interests exist.

REFERENCES

1.   Coppens P, Delmulle L, Gulati O, Richardson D, Ruthsatz M, Sievers H, et al. The use of botanicals in food supplements. Regulatory scope, scientific risk assessment and claims substantiation. Ann Nutr Metab. 2006;50:538-554.

2.   Naidu AS, editor. Natural food and antimicrobial systems. Boca Raton (Florida): CRC Press; 2000.

3.   Decker EA, Elias RJ, and McClements DJ, editors. Oxidation in foods and beverages and antioxidant applications. Volume 1: Understanding mechanisms of oxidation and antioxidant activity. Cambridge (UK): Woodhead Publishing; 2010.

4.   Medina I, González MJ, Pazos M, Della Medaglia D, Sacchi R, and Gallardo JM. Activity of plant extracts for preserving functional food containing n-3-PUFA. Eur Food Res Technol. 2003;217(4):301-307.

5.   Pazos M, Gallardo JM, Torres JL, Medina I. Activity of grape polyphenols as inhibitors of the oxidation of fish lipids and frozen fish muscle. Food Chem. 2005; 92(3):547-557.

6.   Moure A, Cruz JM, Franco D, Domínguez JM, Sineiro J, Domínguez H, et al. Natural antioxidants from residual sources. Food Chem. 2001;72(2):145-171.

7.   Bates MP, Phillips PS. Sustainable waste management in the food and drink industry. Brit Food J. 1999;101(8):580-589.

8.   Shoji T, Akazome Y, Kanda T, Ikeda M. The toxicology and safety of apple polyphenol extract. Food Chem Toxicol. 2004;42(6):959-967.

9.   Yamakoshi J, Saito M, Kataoka S, Kikuchi M. Safety evaluation of proanthocyanidin-rich extract from grape seeds. Food Chem Toxicol. 2002;40(5):599-607.

10. Gilsenan MB. Nutrition & health claims in the European Union: a regulatory overview. Trends Food Sci Technol. 2011;22:536-542.

11. Sun JIE, Chu Y-F, Wu X, Liu RH. Antioxidant and antiproliferative activities of common fruits. J Agr Food Chem. 2002;50(25):7449-7454.

12. Boyer J and Liu RH. Apple phytochemicals and their health benefits. Nutrition Journal. 2004;3:5.

13. Khanizadeh S, Tsao R, Rekika D, Yang R, Charles MT, Rupasinghe VHP. Polyphenol composition and total antioxidant capacity of selected apple genotypes for processing. J Food Compost Analysis. 2008;21(5):396-401.

14. Peschel W, Sánchez-Rabaneda F, Diekmann W, Plescher A, Gartzía I, Jiménez D, et al. An industrial approach in the search of natural antioxidants from vegetable and fruit wastes. Food Chem. 2006;97(1):137-150.

15. Huber GM, Rupasinghe HPV. Phenolic profiles and antioxidant properties of apple skin extracts. J Food Sci. 2009;74(9):C693-C700.

16. Spigno G, Tramelli L, De Faveri DM. Effects of extraction time, temperature and solvent on concentration and antioxidant activity of grape marc phenolics. J Food Eng. 2007;81(1):200-208.

17. Ciurzyńska A, Lenart A. Freeze-drying: application in food processing and biotechnology - A review. Polish J Food Nutr Sci. 2011;61(3):165-171.

18. Henríquez C, Speisky H, Chiffelle I, Valenzuela T, Araya M, Simpson R, et al. Development of an ingredient containing apple peel, as a source of polyphenols and dietary fiber. J Food Sci. 2010;75(6):172-181.

19. Rupasinghe HPV, Wang L, Huber GM, Pitts NL. Effect of baking on dietary fibre and phenolics of muffins incorporated with apple skin powder. Food Chem. 2008; 107(3):1217-1224.

20. Wolfe K, Liu RH. Apple peels as a value-added food ingredient. J Agr Food Chem. 2003;51:1676-1683.

21. Wegrzyn TF, Farr JM, Hunter DC, Au J, Wohlers MW, Skinner MA, et al. Stability of antioxidants in an apple polyphenol-milk model system. Food Chem. 2008;109(2):310-318.

22. Tanabe M, Kanda T, Yanagida A. Process for the production of fruit polyphenols from unripe Rosaceae fruit. The Nikka Whisky Distilling Co. Ltd., Tokyo (Japan). U.S. Patent. 1994;5:932,623.

23. Wolfe K, Wu X, Liu RH. Antioxidant activity of apple peels. J Agr Food Chem. 2003;51(3):609-614.

24. Carter S, Shaw SA. The UK apple industry: production, markets and the role of differentiation. Brit Food J. 1993;95(10):23-28.

25. Bord Glas. National Apple Orchard Census 2002. 2003; Accessed 14 March 2007. Available: http://www.bordglas.ie.

26. Bord Bia. National Apple Orchard Census 2007. 2008; Accessed 3 May 2012. Available: http://www.bordbia.ie.

27. Imeh U, Khokhar S. Distribution of conjugated and free phenols in fruits: antioxidant activity and cultivar variations. J Agr Food Chem. 2002;50:6301-6306.

28. Price KR, Prosser T, Richetin AMF, and Rhodes MJC. A comparison of the flavonol content and composition in dessert, cooking and cider-making apples; distribution within the fruit and effect of juicing. Food Chem. 1999;66(4):489-494.

29. McCann MJ, Gill CIR, O' Brien G, Rao JR, McRoberts WC, Hughes P, et al. Anti-cancer properties of phenolics from apple waste on colon carcinogenesis in vitro. Food Chem Toxicol. 2007;45(7):1224-1230.

30. Massini L, Martin Diana AB, Barry-Ryan C, and Rico D. Apple (*Malus domestica* Borkh. cv Bramley's Seedling) peel waste as a valuable source of natural phenolic antioxidants. In: Waldron KW, Moates GK, and Faulds CB, editors. Total Food - Sustainability of the Agri-Food Chain. Cambridge: RCS; 2010.

31. AOAC. Official method of analysis. 17th ed. Gaithersburg: AOAC Int.; 2000.

32. Prosky L, Asp NG, Furda I, De Vries JW, Schweizer TF, Harland B. The determination of total dietary fiber in foods, food products: collaboratory study. J AOAC Int. 1985;68:677-679.

33. Dubois M, Gilles KA, Hamilton KA, Rebers PA, Smith F. Colorimetric method for determination of sugars and related substances. Anal Chem. 1956;28:350-356.

34. Singleton VL, Orthofer R, Lamuela-Raventos RM. Analysis of total phenols and other oxidation substrates and antioxidants by means of Folin-Ciocalteu reagent. Method Enzymol. 1999;299:152-178.

35. Zhishen J, Mengcheng T, Jianming W. The determination of flavonoid contents in mulberry and their scavenging effects on superoxide radicals. Food Chem. 1999;64(4):555-559.

36. Stratil P, Klejdus B, Kubáň V. Determination of total content of phenolic compounds and their antioxidant activity in vegetables - Evaluation of spectrophotometric methods. J Agr Food Chem. 2006;54(3):607-616.

37. Makris DP, Boskou G, Andrikopoulos NK. Polyphenolic content and in vitro antioxidant characteristics of wine industry and other agri-food solid waste extracts. J Food Compost Analysis. 2007;20(2):125-132.

38. Brand-Williams W, Cuvelier ME, Berset C. Use of a free radical method to evaluate antioxidant activity. Lebensm Wiss Technol, 1995; 28: 25-30.

39. Almeida-Doria RF, Regitano-d'Arce MAB. Antioxidant activity of rosemary and oregano ethanol extracts in soybean oil under thermal oxidation. Ciênc Tecnol Aliment. 2000;20:197-203.

40. Łata B. Relationship between apple peel and the whole fruit antioxidant content: year and cultivar variation. J Agr Food Chem. 2007;55(3):663-671.

41. Chinnici F, Bendini A, Gaiani A, Riponi C. Radical scavenging activities of peels and pulps from cv. Golden Delicious apples as related to their phenolic composition. J Agr Food Chem. 2004;52(15):4684-4689.

42. Vinson JA, Su X, Zubik L, Bose P. Phenol antioxidant quantity and quality in foods: fruits. J Agr Food Chem. 2001;49(11):5315-5321.

43. Guyot S, Marnet N, Laraba D, Sanoner P, Drilleau J-F. Reversed-Phase HPLC following thiolysis for quantitative estimation and characterization of the four main classes of phenolic compounds in different tissue zones of a French cider apple variety (Malus domestica var. Kermerrien). J Agr Food Chem. 1998;46(5):1698-1705.

44. Tsao R, Yang R, Xie S, Sockovie E, Khanizadeh S. Which polyphenolic compounds contribute to the total antioxidant activities of apple? J Agr Food Chem. 2005;53(12):4989-4995.

45. Apak R, Güçlü K, Demirata B, Özyürek M, Çelik SE, Bektaşoğlu B, et al. Comparative evaluation of various total antioxidant capacity assays applied to phenolic compounds with the CUPRAC assay. Molecules. 2007;12(7):1496-1547.

46. Foti MC, Daquino C, Geraci C. Electron-Transfer reaction of cinnamic acids and their methyl esters with the DPPH• radical in alcoholic solutions. J Org Chem. 2004; 69(7):2309-2314.

47. Prior RL, Wu X, Schaich K. Standardized methods for the determination of antioxidant capacity and phenolics in foods and dietary supplements. J Agr Food Chem. 2005;53(10):4290-4302.

48. Kondo S, Tsuda K, Muto N, Ueda J-E. Antioxidative activity of apple skin or flesh extracts associated with fruit development on selected apple cultivars. Sci Hortic, 2002;96(1-4):177-185.

49. Koşar M, Dorman HJD, Hiltunen R. Effect of an acid treatment on the phytochemical and antioxidant characteristics of extracts from selected Lamiaceae species. Food Chem. 2005;91(3):525-533.

Scientific Opinion on the Regulatory Status of 1,3-Dimethylamylamine (DMAA)

Bastiaan J. Venhuis[1*] and Dries de Kaste[1]

[1]RIVM - National Institute for Public Health and the Environment, P.O. Box 1, NL-3720 BA Bilthoven, The Netherlands.

Authors' contributions

This work was carried out in collaboration between all authors. Author BJV performed the literature study, data analysis and wrote the first draft of the manuscript. Author DK took the initiative to prepare a scientific opinion and performed a critical review of the first draft. All authors read and approved the final manuscript.

SUMMARY

1,3-Dimethylamylamine (DMAA) is a pressor amine often found in food supplements for athletes at dosages of 25-65 mg. Historically, the compound has been used as a nasal decongestant but its oral application is largely unstudied leaving the regulatory status of such food supplements as unlicensed medicines undetermined. We therefore reviewed the literature on DMAA and similar amines in order to deduce an effective oral dosage. Based on our findings we conclude that oral preparations with >4 mg DMAA per dose unit should be considered as effective as a bronchodilator. Food supplements that exceed that limit are in fact subject to the Medicines Act and require licensing. Dosages higher than 100-200 mg are expected to cause serious adverse events.

Keywords: 1,3-Dimethylamylamine (DMAA); geranamine; regulatory status; food supplements; oral efficacy.

1. INTRODUCTION

DMAA is an aliphatic amine with vasoconstricting properties. It can easily be prepared synthetically [1,2] and was used between 1940-1970 as the active pharmaceutical ingredient in Forthane® nasal inhalers (Lilly) for relieving nasal congestion. Although DMAA is barned

Corresponding author: Email: Bastiaan.Venhuis@rivm.nl

by the international anti-doping organisation (WADA) as a stimulant it is often listed on the label of food supplements for athletes. Laboratory analysis showed that products may contain 25-65 mg DMAA per dose unit (data not shown) [3,4]. Zhang et al. (2012) even found dosages of 285 mg [5].

According to EU law the marketing of pharmacologically active compounds for human consumption requires licensing as a medicine. To alleviate regulatory pressure on natural products this definition is applied to food supplements only when they are pharmacological effective. Since the oral efficacy of DMAA is poorly understood this left its regulatory status as an oral medicine undetermined [5].

To determine whether food supplements with DMAA should be considered as unlicensed medicines we reviewed international scientific and patent literature on the pharmacology of DMAA and similar compounds using Scopus, Pubmed and SureChem. Based on our findings we formulate a scientific opinion on the regulatory status of DMAA in food supplements.

2. NOMENCLATURE

DMAA is one of the many trivial names for '4-methyl-hexane-2-amine' (Table 1). The trivial name 'Geranamine' refers to geranium oil being claimed as a natural source of DMAA [6]. However, elaborate studies could not confirm the presence of DMAA in geranium oil and even suggest that this claim may have been fabricated in order to justify its use in food supplements [7,8].

Table 1. Systematic name and trivial names for DMAA

4-Methyl-hexane-2-amine (syst. name)	Forthan
Geranamine	Methylhexaneamine
1,3-Dimethylpentylamine	4-Methyl-2-hexylamine
Floradrene	4-Methyl-2-hexaneamine
Forthane*	Forthan

*The brand name 'Forthane' is currently attached to an anesthetic unrelated to DMAA.

The molecular structure of DMAA contains two chiral centres (Fig. 1). For reasons of clarity this document only uses the name DMAA for all optical isomers and mixtures thereof, free base and all salt forms unless stated otherwise.

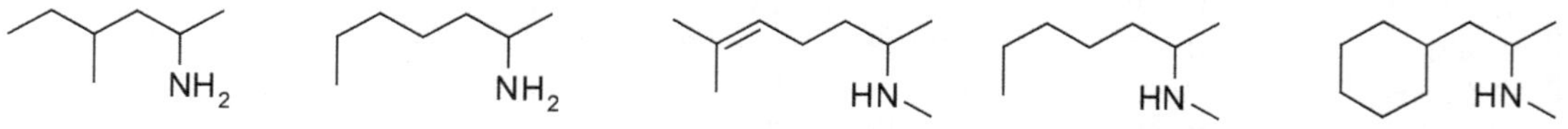

Geranamine (Lilly)   Tuaminoheptane (Lilly)   Octin (Bilhuber)   Oenethyl (Bilhuber)   Propylhexedrine (Smith-Kline & French)

Fig. 1. Molecular structures of aliphatic amines that were used in medicines.

3. PHARMACOLOGY

DMAA is an indirect sympathomimetic with vasoconstricting properties and cardiovascular effects [9]. These properties are common among many other aliphatic amines which therefore are generally referred to as 'pressor amines' [10-14]. Table 2 shows the pharmacological effects of DMAA observed in several species of animals.

Table 2. Pharmacological effects of DMAA in animal studies

System	Effects	Species
Cardiovascular	Increase in arterial blood pressure	Dog, cat, rat [9,12,15]
	Vasoconstriction	Rat, frog [16]
	Myocardial depression	Rat, guinea pig, rabbit, dog [16]
	Tachycardia	Dog [9]
Respiratory	Bronchodilation, increased nasal and lung volume	Dog [9]
CNS	Shortening of pentobarbital narcosis	Mouse [16]
Renal system	Diuresis	Dog [9]
Intestines	Depression of the peristaltic activity	Guinea pig, rat [16]
Reproductive organs	Antagonises acetylcholine induced contraction of the uterus	Rat [16]

Animal studies with DMAA have shown some similarities with the effects of ephedrine and amphetamine [9,11,12,17]. Literature shows that DMAA is largely equipotent to tuaminoheptane on most pharmacological aspects [16,17]. The hypertensive effect of DMAA was best described in literature. This suggested that the hypertensive effect of DMAA (iv) is 1-2x the effect for the same dose of tyramine [15]. In the dog, the hypertensive effect after oral dosing showed to be 4x weaker than ephedrine and 2x weaker than amphetamine [17]. Toxicological data are scarce. DMAA did not induce eye irritation in the rabit and the LD_{50} values (iv) were 39.0 and 72.5 mg/kg in mice and rats, respectively [16].

The therapeutic potential of aliphatic amines was extensively investigated by pharmaceutical industry in the first part of the 20th century [9,11-13,17]. At least 5 different aliphatic amines were eventually marketed as an active ingredient of a medicine: DMAA, tuaminoheptane, octin, oenethyl and propylhexedrine (Table 3) [11]. The therapeutic use of DMAA was limited to a nasal decongestant. Tuaminoheptane, octin en oenethyl were also used as medicines for their cardiovascular, bronchodilating, or anti-migraine properties. Propylhexedrine was also used as a weight-loss drug [23]. Detailed clinical studies were not found in scientific literature. DMAA and all structurally similar compounds are banned as stimulants by the World Anti-Doping Authority (WADA).

4. INTRANASAL APPLICATION

In 1944-1945 patents were granted to Lilly on the medical use of certain aliphatic amines, specifically for DMAA [1,2]. The patents describe the vasoconstricting action of DMAA and its usefulness as a nasal decongestant. DMAA was marketed as a nasal decongestant by Lilly under the name of 'Forthane®' until about 1970 [24]. Lilly advertisements for Forthane state that intranasal use should normally not cause psychic disturbances, rise in blood pressure, cardiac irregularities or over-constriction [24].

Table 3. Aliphatic amines with properties similar to DMAA

Compound	Description
Tuaminoheptane (also: Tuamine)	Initially used as a nasal decongestant. An inhaler contained 325 mg tuaminoheptane adsorbed on a cotton plug [18]. The dosage of the spray was about 0.25 mg per spray [19].
Octin (also: Isometheptene)	Initially marketed as a nasal decongestant but still in use to control blood pressure during anesthesia, and for the treatment of migraine and muscle cramps [14,20].
Oenethyl	Initially marketed as a nasal decongestant and also used to control blood pressure during anesthesia [21,22]. ENREF_23
Propylhexedrine	Marketed as a nasal decongestant and also used as a weight-loss drug with a low potential for abuse [23].

Forthane® was a multiple use nasal inhaler containing DMAA carbonate adsorbed a cotton plug (equivalent to 250 mg DMAA as a free base). An effective dose is not reported for this drug delivery system. However, the Lilly patent also reports a different formulation; an effective nasal spray with a concentration of 0.6 mg/ml. Assuming the spray volume is not more than 1 ml, intranasal dosages of 0.6 mg and higher should be considered as therapeutically effective.

5. ORAL APPLICATION

Scientific literature was searched for articles mentioning the oral application of DMAA and similar amines. Only three articles were retrieved that described the oral use of DMAA. Several other articles were retrieved describing the effects of DMAA in laboratory animals and of similar amines in laboratory animals and humans.

Marsh et al. [25] described that a single oral dose of 3 mg/kg in a human (210 mg/70 kg) results in a moderate increase in the heart rhythm and blood pressure. In addition to a dry mouth, a runny nose and goose pimples, the subject also experienced serious side effects such as confusion and concentration problems. No signs of stimulation could be identified in the central nervous system. Lower doses were not investigated.

Perrenoud et al. [3] described the excretion of DMAA in urine following a single oral dose of 40 mg. They found that about 32 mg DMAA was found unchanged in the urine. The authors did not investigate the pharmacological effects.

Lisi et al. [7] described the excretion of DMAA in urine following a single oral dose of an undetermined quantity.

6. DISCUSSION

The pharmacology of DMAA is best known after intranasal application for relieving nasal congestion (Forthane®). Its medicinal use is based on the vasoconstricting effects in the nose after local application. Nevertheless, literature shows that systemic exposure may cause vasoconstriction of blood vessels of the lungs and the heart. The article by Perrenoud et al. shows that DMAA is absorbed very well by the body when taken orally and that it is hardly metabolized [3]. DMAA is absorbed relatively slowly by the body (4-12 hours) and urinary excretion is very slow ($t_{0,5}$ = 24 hours). The slow urinary excretion of unchanged DMAA is also shown by Lisi et al. [7]. Therefore, there is a risk of a buildup of doses through daily use whereby the efficacy increases.

Because scientific literature on the oral use of DMAA is scarce its oral efficacy was deduced by assessing its intrinsic efficacy and oral efficacy relative to known values for similar amines (Table 4).

Table 4. Literature data on the pharmacological effect of DMAA and similar amines on the increase of blood pressure in dogs and humans

Blood pressure increase in dogs (iv dosing)		
Authors	**Compound**	**Normalization to epinephrine (rel. intrinsic efficacy)**
Swanson (1946)[12]	DMAA sulfate	4,7 mmol DMAA ≈ 3.55 µg epinephrine (1 : 0.76)
Marsh (1951) [25]	DMAA HCl	4,6 mmol DMAA ≈ 3.7 µg epinephrine (1: 0.80)
Swanson (1948) [17]	Propylpentadrine HCl*	1 mmol DMAA ≈ 0.97 µg epinephrine (1 : 0.97)
	Propylhexedrine HCl	1 mmol DMAA ≈ 0.26 µg epinephrine (1 : 0.26)

Blood pressure increase in dogs (po dosing)		
Author	**Compound**	**Rel. oral efficacy**
Swanson (1948) [17]	DMAA sulfate	0.93 – 1.88 mmol = 20% effect
	Propylpentadrine HCl	0.56 – 0.84 mmol = 100% effect
	Ephedrine sulfate	0.63 – 1.25 mmol = 90% effect

Blood pressure increase (17-23 mmHg) in humans (po dosing)		
Author	**Compound**	**Dose**
Marsh (1949) [26]	Ephedrine HCl	25 mg
	Propylhexedrine HCl	120 mg

Relative efficacy iv (intrinsic efficacy) versus po		
Compound	**Rel. intrinsic efficacy**	**Rel. oral efficacy**
DMAA	0.8	0.2
Propylpentadrine	0.97	1
Propylhexedrine	0.26	0.26[#]
Ephedrine	-	0.9

Propylpentadrine = 2-Methylamino-1-cyclopentyl-propane
 # Assumption by the authors based on structural similarity with Propylpentadrine.

6.1 Effect on the Heart

Propylhexedrine (Benzedrex®, B.F Ascher) is a DMAA analogue that is used as an inhaler in the United States to treat nasal congestion. Old and modern inhalers give off 0.4-0.5 mg propylhexedrine per 800 ml of inhaled air [27,28]. Propylhexedrine has a therapeutic effect on, for example, bradycardia [29] both after inhalation (dose not reported) as well as after oral application (50 mg). The intrinsic efficacy of DMAA measured against the effect of blood pressure is about 3 fold higher than that of propylhexedrine, but this is compensated by a 4 fold poorer oral absorption [3,17]. Therefore, effects on the heart similar to an oral dose of 50 mg benzedrex may be expected after a oral dose of about 50-75 mg DMAA.

6.2 Effect on the Blood Pressure

The effect of orally-dosed DMAA on the blood pressure of dogs is 4.5 times lower than that of orally-dosed ephedrine [17]. After intravenous application, the relative effect of DMAA is actually 4 times higher. This confirms a comparable intrinsic working and that a slow oral absorption of DMAA leads to a lower peak dose ($C_{max.}$). With ephedrine, the blood pressure increasing effects have been seen in humans after a single oral dose of 25-60 mg. [30,31]. The slow oral absorption of DMAA means a blood pressure increasing effect is expected after a single dose of up to about 100 mg.

An oral dose of 97 mg (free base) of propylhexedrine in humans results in an increase in blood pressure of 17-23 mmHg. [26]. It is expected that the greater intrinsic efficacy of DMAA on the blood pressure will be compensated by a slower oral absorption [3,17]. Based on this comparison an increase in blood pressure can also be expected after a single dose of DMAA of about 100 mg.

6.3 Effect on the Lungs and the Nose

Ephedrine was applied orally to alleviate nasal congestion and as a bronchodilator in doses of 15-60 mg. [31]. It does not seem to work as a bronchodilator at lower doses. As the free base DMAA is 19 times as potent as ephedrine in alleviating nasal congestion [32]. However, the oral efficacy of DMAA is 4.5 times less than that of ephedrine. Therefore, a single oral dose of 4-15 mg DMAA can be expected to be equivalent to a single oral dose of 15-60 mg ephedrine.

7. CONCLUSION

According to literature DMAA is a pharmacologically active compound of synthetic origin. The single known reason for using DMAA in humans is pharmacological. The body absorbs DMAA adequately, but relatively slowly, when taken orally. Pharmacological effects after oral intake can be expected on the lungs (bronchodilation) and the nasal mucosa following a singe oral dose of about 4-15 mg. Pharmacological effects on the heart can be expected following a single oral dose of about 50-75 mg. Pharmacological effects on the blood pressure can be expected after a single oral dose of about 100 mg. Because of the long half-life, there is a risk that repeated doses within 24-36 hours could lead to steadily stronger pharmacological effects (build up). Ba sed on our findings we conclude that oral preparations with >4 mg DMAA should be considered as effective as a bronchodilator. Therefore, food supplements containing >4 mg DMAA should be treated as subject to the Medicines Act and require licensing as a medicine. Dosages higher than 100-200 mg are expected to cause serious adverse events.

COMPETING INTERESTS

Authors have declared that no competing interests exist.

REFERENCES

1. Lilly E. Aminoalkanes, Patent US2. 1944;350:318.
2. Lilly E. Carbonates of 1-R-1 aminoethanes. Patent US2. 1945;386:273.
3. Perrenoud L, Saugy M, Saudan C. Detection in urine of 4-methyl-2-hexaneamine, a doping agent. J Chromatogr B Analyt Technol Biomed Life Sci. 2009;877:3767-70.
4. Perrenoud et al. found 38 mg DMAA per dose unit in a food supplement. Using their analysis method our laboratory found 25 mg and 65 mg per dose unit in two other food supplements. The label instruction of the first advised to take 1-3 doses daily (25-75 mg).
5. Zhang Y, Woods RM, Breitbach ZS, Armstrong DW. 1,3-Dimethylamylamine (DMAA) in supplements and geranium products: Natural or synthetic? Drug Testing and Analysis; 2012.
6. Ping Z, Jun Q, Qing L. A Study on the chemical constituents of geranium oil. Journal of Guizhou Institute of Technology. 1996;25:82-85.
7. Lisi A, Hasick N, Kazlauskas R, Goebel C. Studies of methylhexaneamine in supplements and geranium oil. Drug Testing and Analysis. 2011; 3:873-876.
8. Zhang Y, Woods RM, Armstrong DW. 1,3-Dimethylamylamine (DMAA) in supplements and geranium plants/products: natural or synthetic? Drug Testing and Analysis. 2012;4.
9. Charlier R. Pharmacology of 2-amino-4-methylhexane. Arch Int Pharmacodyn Ther. 1950;83:573-84.
10. Barger G, Dale HH. Chemical structure and sympathomimetic action of amines. J Physiol. 1910;41:19-59.
11. Ahlquist RP. A contribution to the pharmacology of the aliphatic amines. Journal of Pharmacology and Experimental Therapeutics. 1944;81:235-239.
12. Swanson EE, Chen KK. Comprison of pressor action of aliphatic amines. Journal of Pharmacology and Experimental Therapeutics. 1946;88:10-13.
13. Ahlquist RP. The sympathomimetic vasodilating action of the aliphatic amines'. Journal of Pharmacology and Experimental Therapeutics. 1945;85:283-287.
14. Fellows EJ. The pharmacology of 2-amino-6-methylheptane'. Journal of Pharmacology and Experimental Therapeutics. 1947;90:351-358.
15. Holtz P, Palm D. On the mechanism of sympathomimetic action of some aliphatic amines. Naunyn Schmiedebergs Arch Exp Pathol Pharmakol. 1965;252:144-58.
16. Miya TS, Edwards LD. A pharmacological study of certain alkoxyalkylamines. J Am Pharm Assoc Am Pharm Assoc (Baltim). 1953;42:107-10.
17. Swanson EE, Chen KK. Comparison of pressor action of alicyclic derivatives of aliphatic amines. J Pharmacol Exp Ther. 1948;93:423-9.
18. Comer JP, Kennedy EE. Assay of 2-aminoheptane and methyl hexane amine in plastic inhalers by titration in glacial acetic acid. Journal of the American Pharmaceutical Association. 1956;45:454-455.
19. Thevis M, Sigmund G, Koch A, Schanzer W. Determination of tuaminoheptane in doping control urine samples. Eur J Mass Spectrom (Chichester, Eng). 2007;13:213-21.
20. Gruber CM, Heiligman R, Denote A. The effect of methyl-amino methyl heptene (Octin) upon the intact intestine in the non-anesthetized dog. Journal of Pharmacology and Experimental Therapeutics. 1936;56:284-289.
21. Booker WM, Molano PA, French DM. Effect of oenethyl on respiration and on blood pressure in anesthetized dogs. Curr Res Anesth Analg. 1949;28:121-9.
22. Zeldis N. Oenethyl in spinal anesthesia., Curr Res Anesth Analg. 1953;32:27-36.
23. Sweetman SC. Martindale, The complete Drug Reference; 36th ed.; Pharmaceutical Press, London; 2009.

24. Lilly E. NEW and nonofficial remedies: methylhexamine; forthane. J Am Med Assoc. 1950;143:1156.
25. Marsh DF, Howard A, Herring DA. The comparative pharmacology of the isomeric nitrogen methyl substituted heptylamines. Journal of Pharmacology and Experimental Therapeutics. 1951;103:325-329.
26. Marsh DF, Herring DA. The comparative pressor activity of the optical isomers of cyclohexylisopropylamine and cyclohexylisopropylmethylamine. J Pharmacol Exp Ther. 1949;97:68-71.
27. Burnett JB, Gundersen SM. The effect of volatile L-Cyclohexyl-2-Methylaminopropane (Benzedrex Inhaler) on patients with coronary arteriosclerosis. New England Journal of Medicine. 1952;246:449-450.
28. B.F. Ascher Consumer Health Care Products, www.bfascher.com/ourproducts.html.
29. Day J, Viar WN. A case of heart block treated with 1-cyclohexyl-2-methylaminopropane (benzedrex). American Heart Journal. 1951;42:733-736.
30. Haller CA, Jacob P, 3rd; Benowitz NL. Enhanced stimulant and metabolic effects of combined ephedrine and caffeine. Clin Pharmacol Ther. 2004;75:259-73.
31. Drew CD, Knight GT, Hughes DT, Bush M. Comparison of the effects of D-(-)-ephedrine and L-(+)-pseudoephedrine on the cardiovascular and respiratory systems in man'. Br J Clin Pharmacol. 1978;6:221-5.
32. Proetz AW. Certain aliphatic componds as nasal vasoconstrictors. Archives of Otolaryngology--Head & Neck Surgery. 1943;37:15-22.

Isolation of Lactic Acid Bacteria from Ewe Milk, Traditional Yoghurt and Sour Buttermilk in Iran

Mahdieh Iranmanesh[1], Hamid Ezzatpanah[1*], Naheed Mojgani[2],
Mohammad Amir Karimi Torshizi[3], Mehdi Aminafshar[4]
and Mohamad Maohamadi[1]

[1]Department of Food Science and Technology, Science and Research
Branch, Islamic Azad University, P.O. Box 14515.775, Tehran-Iran.
[2]Biotechnology Department, Razi Vaccine and Serum Research Institute, Karaj-Iran.
[3]Department of Poultry Science, Faculty of Agriculture, Tarbiat Modares University,
Tehran-Iran.
[4]Department of Animal Science, Science and Research Branch, Islamic Azad University,
P.O. Box14515.775, Tehran-Iran.

Authors' contributions

This work was carried out in collaboration between all authors. MI designed the study, MA performed the statistical analysis, MI, HE, NM wrote the protocol, HE, MI wrote the first draft of the manuscript. MAKT, HE, NM managed the analyses of the study. MI, HE, NM managed the literature searches. MM managed the sampling. All authors read and approved the final manuscript.

ABSTRACT

A total of 63 samples including ewe milk, yoghurt and traditional buttermilk were collected from Myaneh and Hashrood (Azarbayjan-e-Sharqi, Iran) and screened for the presence of Lactic Acid Bacteria (LAB). Based on routine cultural characteristic, general morphological and biochemical assay, 77 out of 168 bacterial isolates were identified as LAB. These isolates were examined for the presence of inhibitory activity against other randomly selected LAB isolates. Thirty-three strains showed antagonistic activity against the closely related LAB strains and were further challenged against other gram-positive and gram-negative pathogens including *Listeria monocytogenes*, *Staphylococcus aureus* and *Salmonella entritidis*. Based on their zones of inhibition diameters the isolates showing maximum inhibitory activity against these pathogens were selected for detailed

Corresponding author: Email: hamidezzatpanah@srbiau.ac.ir

investigations. The selected isolates were identified to species level by 50CHL API system and were challenged to heat, acid and bile salt. Most of the strains were able to survive at different pH ranges, while one strain of *Pedicoccus acidilactici* and *Lactobacillus paracasei* were able to tolerate all ranges of pH during 24 h of incubation. In addition, *Lactobacillus brevis* was found as the most resistant strain being able to resist all concentrates of bile after 4 h. The results indicated the probiotic potential of the isolates, as majority of the selected LAB isolates were capable of resisting high temperatures, acidic pH values and bile concentrations of 0.7%.

Keywords: Traditional buttermilk; lactic acid bacteria; antimicrobial activity; probiotic bacteria.

1. INTRODUCTION

A number of reports have emphasized the significance of food fermentation mainly because of degradation or inactivation of anti-nutritive factors, toxins, as well as improvement of digestibility of foods that leads their major role in the diet of different regions (Mathara et al., 2008). Although, different components of milk are involved in fermentation process, LAB commonly metabolize lactose into lactic acid resulting in pH reduction and higher titratable acidity (TA) and creating an environment that is unfavorable to pathogens and spoilage organisms (Aslim et al., 2005).

According to reports, it appears that Middle East is the origin of fermented dairy products mainly yoghurt (Tamime and Robinson, 2007). In Iran, a number of traditional dairy products are consumed of which yoghurt, well known, as Mast is one of the most popular fermented milk products. While, traditionally made sour buttermilk especially made from ewe milk is more common in rural areas of the country.

Several previously published reports have indicated the presence of *Lactobacillus* strains in sheep and cow milk (Mobarez et al., 2008). In addition, several studies have shown the inhibitory activities of numbers of LAB such as *Lactobacillus brevis* isolated from Turkish dairy products (Aslim et al., 2005) and *Lactobacillus acidophilus* isolated from Iranian yoghurt against *Staphylococcus aureus* (Mobarez et al., 2008).

It is a well-established fact that the composition of LAB in these traditional dairy products is varied and inconstant. Owing to the said health benefit of buttermilk in the rural areas it appeared interesting to evaluate the microbial load of these products with specific emphasis on probiotic bacteria mainly LAB. The main objective of present study was to investigate LAB of traditional sour buttermilk made from ewe's milk, which might provide important information regarding its probiotic potential and its utilization in the future.

2. MATERIALS AND METHODS

2.1 Bacterial Strains and Culture Conditions

All LAB strains used in this study were grown in MRS broth (HiMedia India) at 37°C for 24-48 h in anaerobic jars. All pathogenic strains used in this study were grown in BHI (HiMedia-India) at 37°C for 18-24 h under aerobic condition.

All strains were maintained at 4°C (in aerobic condition) and renewed every week for short-term preservation. The long-term conservation of the purified isolates was carried out in MRS broth with sterile glycerol (15%) and stored at -70ºC (Badis et al., 2004).

2.2 Collection of Ewe Milk and Preparation of Buttermilk

Ewe milk samples were collected from 30 sheep herds in Myaneh (15 herds) and Hashrood (15 herds) cities in north-west of Iran. All samples were collected according to ISO 707 in sterile bottles of 250 mL and transported to the laboratory under refrigeration (4ºC) within 36 h (AOAC, 2002).

The traditionally made yoghurt and sour buttermilk samples were also collected from the same area. Fig.1 shows the preparation operation of yoghurt and sour buttermilk. As indicated, the samples are prepared in a traditional device known as Mashk (Fig. 2) that is made from hide (sheepskin) and is used for making butter and buttermilk from yoghurt. This device is also used widely for preservation of fermented dairy products for 10 to 20 days at temperatures not exceeding 20ºC.

2.3 Isolation and Identification of LAB

One ml of the collected samples (ewe milk, yoghurt and sour buttermilk) were inoculated in MRS broth and incubated at mentioned conditions until the appearance of growth. A loop full of the grown culture broth was then plated on MRS agar plates and pure colonies selected after incubation and tested for Gram identification, cell morphology and catalase activity (Karimi Torshizi et al., 2008).

2.4 Screening for Antagonistic Interactions among LAB Strains

The antagonistic effects of isolated LAB against four other selected LAB isolates (which showed the fastest growing strains and retained their maximum viability during storage at refrigerated temperatures) were determined by using an agar diffusion method (Aslim et al., 2005) with slight modifications. The freshly grown overnight cultures of the producer and indicator strains were adjusted to McFarland Index 3 prior to use. The surface of MRS agar plates were evenly streaked with 0.1 ml of a 24h broth culture of the selected indicator strains, with a sterile cotton swab. The culture broth of the producer strains (150ul) were poured into the wells (10mm) made in these agar plates with a sterile borer (2.5mm). All plates were stored for 2h at 4ºC prior to incubation at 37ºC for 24h. The antimicrobial activity was recorded as appearance of clear zone around the wells and the zone diameter (+:3mm<zone, ++:3mm<zone<5mm, +++:5mm<zone<7mm) measured in millimeter. All tests were run in duplicate.

2.5 Antimicrobial Activity against Pathogens

Thirty-three LAB isolates demonstrating high antibacterial activity against other selected LAB isolates were further checked against other Gram positive and negative pathogens by agar well diffusion method described earlier. *Staphylococcus aureus* (PTCC 1112), *Listeria monocytogenes* (PTCC 1298), and *Salmonella enteritidis* (local isolate) were used as indicator culture. As mentioned earlier, the culture broths of both the producer and indicator strains were adjusted to McFarland Index 3 prior to use.

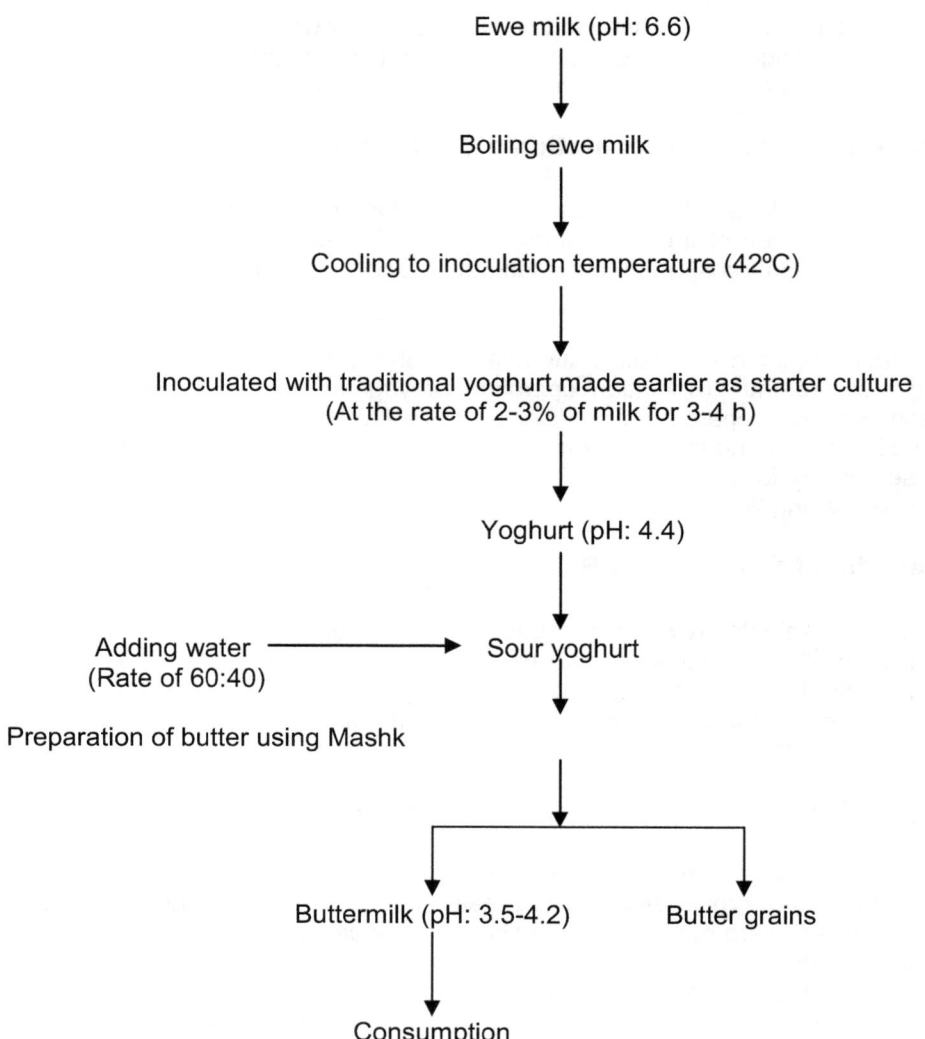

Fig. 1. Preparation of traditional yoghurt and sour buttermilk.

2.6 Identification of Isolated LAB to Species Level

The 10 LAB strains showing maximum activity against the tested pathogens were identified to species level and their carbohydrate fermentation profiles were investigated using API 50 CHL medium (Bio-Meriux, France) according to the manufacture's instruction (Yukeskdag et al., 2004).

2.7 Heat Tolerance

Heat resistance of identified LAB strains was tested by exposing the overnight grown cultures to 55°C and 80°C for 3 min according to the method described earlier (Christiansen et al., 2006). The growth was recorded by inoculating 100ul of the treated cultures on MRS agar plates.

2.8 Acid Tolerance

Acid tolerance of the selected LAB isolates was determined by the method described by Liong and Shah (2005), with slight modifications. The pH of the MRS broth were adjusted to 2.0, 3.0, 4.0, 5.0, and 6.0 by adding 1N HCl. Overnight grown culture broths of the selected LAB isolates were centrifuged and pellet collected, washed twice with PBS and adjusted to Mcfarland 3 Index. Approximately 100ul of the cell culture was inoculated in these tubes and their growth recorded after 0, 1, 2, 4, 24 h of incubation at 37°C. The cell count (cfu/ml) at 0 h was considered as control.

2.9 Bile Tolerance

The effect of bile acid tolerance on the growth of identified LAB strains was determined according to the method described by Gilliland (1984) with slight modifications. The pH of all MRS broth was adjusted to 6.5 and then, 20ul of freshly grown culture of each strains were added to MRS broth with different levels (0.1, 0.3, 0.5, .07, 1, 2, 3 % w/v) of bile salts (oxgall). The bacterial growth monitored after 0, 2, 4 and 24 h of incubation at 37°C by colony count. The cell count (cfu/ml) at 0 h was considered as control.

3. RESULTS

Seventy-seven out of 168 bacterial colonies from 63 samples of ewe milk, yoghurt and sour buttermilk (Table 1) were identified as belonging to genus LAB based on their gram reaction, morphology and catalase test. All isolates were catalase negative and Gram-positive cocci (n=30), bacilli (n=38) and cocobacilli (n=9).

3.1 Antagonistic Reaction

3.1.1 Antagonistic interaction against four selected LAB

All of the isolated LAB strains exhibited antibacterial activity when exposed to other selected LAB isolates used as indicator strain. According to the results, majority of the bacteria isolated from milk or yoghurt exhibited activity against indicator bacteria that were isolated from the same products. For instance, most of the bacteria isolated from milk had activity against indicator bacteria isolated from milk. We also observed that most of bacilli appearing isolates exhibited higher antibacterial activity compared to cocci appearing isolates.

3.1.2 Antagonistic activity against pathogens

Among 77 LAB isolates, only those exhibiting highest activity based on their zone of inhibition diameters (ZID) were further screened for their antagonistic activity against other gram-positive and gram-negative pathogens. The selected LAB isolates screened in this study were isolated from milk (24), yoghurt (1) and buttermilk (8) samples. According to their ZID all isolates showed different level of activity against the tested pathogens.

Table 1. Morphological, cultural and physiological characteristics of the LAB isolates

Group	Gram +ve cocci			Gram +ve bacilli			Gram +ve coccobacilli		
	Milk	Yoghurt	Buttermilk	Milk	Yoghurt	Buttermilk	Milk	Yoghurt	Buttermilk
No. of isolates	24	2	4	4	16	18	8	-	1
Spore forming	-	-	-	-	-	-	-	-	-
Catalase test	-	-	-	-	-	-	-	-	-

Table 2 shows that among the tested isolates, twenty-five inhibited the growth of *Salmonella enteritidis* and twenty isolates inhibited *Staphylococcus aureus* and only ten inhibited the growth of *Listeria monocytogenes*. According to results, eight of the 33 LAB isolates could not inhibit the growth of any of the pathogen tested while, 5 LAB isolates inhibited 1 of the pathogens and 10 inhibited 2 of the pathogens. All ten isolates exhibiting antilisterial activity were also inhibitory towards other two pathogens used in study and were thus selected for further detailed investigations.

Table 2. Antimicrobial activity of LAB isolates against other pathogenic strains

No	Isolates	*L. monocytogenes*	*Staph. aureus*	*S. enteritidis*
1	M2	-	+	+
2	M3	++	+++	++
3	M4	-	-	+
4	M5	-	+	+
5	M7	-	-	-
6	M8	-	-	-
7	M9	-	-	-
8	M10	-	-	-
9	M11	+++	++	+++
10	M15	-	-	+
11	M16	-	++	++
12	M17	-	-	+
13	M20	++	++	+
14	M21	+	+	++
15	M22	-	+	+
16	M23	+	++	++
17	M25	-	+	+
18	M26	-	−	+
19	M27	-	+	+
20	M29	+	+++	+
21	M30	++	+	+
22	M33	-	+	+
23	M35	-	+	+
24	M36	+	++	++
25	Y53	-	+	+
26	B.M64	-	-	+++
27	B.M68	-	-	-
28	B.M71	-	+	+
29	B.M73	+++	+	++
30	B.M74	-	-	-
31	B.M75	-	-	-
32	B.M76	-	-	-
33	B.M77	+	+	++

-: no zone of inhibition.
+: 3mm< zone.
++: 3mm< zone <5mm.
+++: 5mm< zone <7mm.
M denotes bacterial isolates from milk while Y and B.M are LAB isolates from yoghurt and buttermilk, respectively.

The selected ten isolates were identified to species level (Table 3) as *Lactobacillus pentosus*, *Lactobacillus paracasei* (two strains), *Lactobacillus brevis*, *Pediococcus acidilactici* (four strains) and *Lactococcus lactis* (two strains).

Table 3. Identification of the selected LAB isolates to species level using standard API 50CH identification kit

No	Sample	Source	Species identified
1	M3	Milk	*Lactobacillus pentosus*
2	M11	Milk	*Pediococcus acidilactici*
3	M20	Milk	*Lactococcus lactis*
4	M21	Milk	*Lactobacillus paracasei*
5	M23	Milk	*Lactococcus lactis*
6	M29	Milk	*Lactobacillus paracasei*
7	M30	Milk	*Lactobacillus brevis*
8	M36	Milk	*Pediococcus acidilactici*
9	BM73	Buttermilk	*Pediococcus acidilactici*
10	BM77	Buttermilk	*Pediococcus acidilactici*

3.2 Resistance to Heat, Acid and Bile

All the selected LAB isolates in study were able to resist 55 and 80°C for 3 min. However, the level of heat resistance differed among the isolates as was evident by their cfu/ml recorded at different time intervals (data not shown).

The effect of acid on the viability of the selected LAB isolates after 1, 4 and 24 h is shown in Table 4. Among all the isolates in study, *L. paracasei* M29, *P. acidilactici* BM77 were the most acid tolerant as they were able to resist acidic pH values of 2.0 even after 24h of incubation. While, *L. lactis* (M20) appeared to be most acid sensitive strains as could not resist pH values ranging from 2 to 5. In addition, two of the isolates namely *L. brevis* (M30) and *L. pentosus* (M3) were not able to grow only at pH value of 2.0, but good growth was observed at higher pH values used in this study. All LAB isolates were able to tolerate pH 6.0.

As shown in Fig. 4, all strains showed variable range of resistance to different concentrations of bile salts during 4 h of incubation. *P. acidilactici* (M36) and *L. lactis* (M20) appeared to be highly bile sensitive as were not able to grow in bile salt concentrations of 0.5% and 0.7% within 4h of incubation, respectively. While, the most resistant strains in this study appeared to be *L. brevis* M30 which showed highest growth in 1% bile after 4 h and *P. acidilactici* BM73 which was also able to resist 0.3% bile even after 24 h. The growth appeared reciprocal to bile salt concentrations, and with an increase in bile concentration, the growth decreased significantly in all the tested LAB isolates.

Table 4. Acid tolerance of selected LAB isolates during different time intervals

No	LAB isolates	pH 2.0			pH 3.0			pH 4.0			pH 5.0			pH 6.0		
		1h	4 h	24 h	1h	4 h	24 h	1h	4 h	24 h	1h	4 h	24 h	1h	4 h	24 h
1	M3	-	-	-	+	+	-	+	+	-	+	+	+	+	+	+
2	M11	+	+	-	+	+	-	+	+	-	+	+	-	+	+	+
3	M20	-	-	-	-	-	-	-	-	-	-	-	-	+	+	+
4	M21	+	+	-	+	+	-	+	+	-	+	+	-	+	+	+
5	M23	+	+	-	+	+	+	+	+	+	+	+	+	+	+	+
6	M29	+	+	+	+	+	+	+	+	+	+	+	+	+	+	+
7	M30	-	-	-	+	+	-	+	+	-	+	+	-	+	+	+
8	M36	+	+	-	+	+	-	+	+	-	+	+	+	+	+	+
9	BM73	+	+	-	+	+	-	+	+	+	+	+	+	+	+	+
10	BM77	+	+	+	+	+	+	+	+	+	+	+	+	+	+	+

+: growth observed
-: No growth observed.

4. DISCUSSION

The traditional fermented dairy products can be possibly a good source of potential probiotic organisms. In Iran, a number of researchers have reported the isolation of LAB from traditional dairy products like doogh (a very popular minted yogurt beverage made by yoghurt and water seasoned with mint and salt), butter, kashk (a thick whitish liquid similar to whey) and cheese (Tajabadi Ebrahimi et al., 2011). However, there are some traditional fermented dairy products in Iran, which have yet not been evaluated for their health benefit, mainly their probiotic properties. Among such fermented dairy products, one is buttermilk, the microbial ecology and beneficial health effects of which has not been reported earlier. These products are famous especially in the rural areas because of their good natural tastes and flavors.

Our results showed that most of the strains isolated from ewe's milk were gram-positive cocci, while bacilli were more dominant in yoghurt and sour buttermilk samples. It seems that the abundance of bacilli in yoghurt and sour buttermilk is due to the starter cultures such as *L. bulgaricus* and *Strep. thermophilus*, which are mainly responsible for yoghurt formation, as well as higher viability of *L. bulgaricus* in acidic environment (Tamime and Robinson, 2007). Although, the optimal pH for growth of *Lactococci* ranges from 6 to 6.5 (Marth and Steele, 2001) but in our studies the pH of sour buttermilk was approximately 3.8 (Mohamadi et al., 2009), and a number of LAB bacteria were viable at this pH value.

As a functional probiotic, anti-pathogen activity is one of the important properties to be considered. Antimicrobial activity of selected strains may be due to the combination of factors including acid, H_2O_2 and bacteriocin like substances (Gilliland et al., 1984). The 10 selected LAB isolates showing higher inhibition against closely related strains were further tested against other pathogens such as *Salmonella enteritidis*, *Staph. aureus*, and *L. monocytogenes*. The pathogens listed have been reported to be of importance in food products (Gurira and Buys, 2005). During our investigations, we observed that majority of the selected LAB strains were able to inhibit the growth of *Salmonella enteritidis* compared to *Staph. aureus* and *L. monocytogenes*. While, ten selected strains were able to inhibit the growth of all three pathogens used in study. In addition, due to high concerns about the presence of *Listeria* in dairy products, especially as this pathogen has the ability to withstand a large variety of environmental conditions such as refrigeration temperatures, only the isolates demonstrating antilisterial activity were selected for further investigations.

Moreover, viability and survival of probiotic bacteria during passage through the stomach is an important parameter to reach the intestine and provide beneficial effect (Chou and Weimer, 1998). According to reports, the pH in stomach is 0.9, with the presence of food the pH increased and reaches up to 3.0. After the ingestion, 2-4 h takes the stomach to become empty (Erkkila and Petaja, 2000). However, some factors like buffering capacity of food, which is a major factor affecting pH, rate of gastric emptying and physiological state of the bacterium, have affect on the survival strains in stomach (Bertazzoni Minelli et al., 2004). In our studies, some of the tested isolates resisted acidic pH values of 2 even for 24 h. The acid resistance of the selected isolates might be exploited for their use as a probiotic as they might withstand gastric stress and survive in high acid food for longer periods without large reduction in numbers.

The next challenge for potential probiotic survivors in gastrointestinal tract is exposure to bile salts in the upper part of the small intestine (Chou and Weimer, 1999). The concentration of human bile ranges from 0.3% to 0.5%, however 0.3% is considered critical concentration for screening for resistant strains (Gilliland et al., 1984). The most resistant strains in this study appeared to be *P. acidilactici* BM73 that not only resisted 1% bile salt for 4 h but also was also able to survive in 0.3% bile for 24 h. According to our results, *P. acidilactici* BM73, *L. lactis* M23 and *L. paracasei* M29 appeared superior to other strains used in study, respectively. These three isolates were also the most acid tolerant strains, while *L. brevis* M30 also being highly bile tolerant was not able to resist acidic pH values. The difference in acid and bile tolerance of one strain from two species within same genus may be due to differences in their cell wall structure (Conway et al., 1987).

Apart from being acid and bile resistant, another important property of some of these isolates was their temperature tolerance. Heat resistance of bacteria is affected by genetic differences among species, the physiological status of the cells and environmental factors such as pH, water activity, salt content and preservatives (De Angelis et al., 2003). It has been reported previously that some strains of *Lactobacillus* can grow at low and high temperatures (below 15ºC and some strains up to 55 ºC), while the thermal death point of *P. acidilactici* has been reported to be 70ºC for 10 min according to Bergey's Manual. Majority of LAB isolates in study were able to resist 80ºC for 3 minutes but their growth rate differed significantly. At 55ºC, all isolates survived within the mentioned time period without any significant effect on their growth rate. It is obvious that the high temperature used for heating milk (boiling) and adding cold water to sour yoghurt, results in an increased bacilli count in sour yoghurt and buttermilk (Tamime and Robinson, 2007). Interestingly, our results showed that all selected strains tolerate 88ºC for 3 min and because we selected these isolated after boiling which is shown in Fig. 1. Therefore, these selected strains might pass boiling process also, it could be possible that these strains lived in the Mashk and added to sour buttermilk after boiling process.

All ten selected strains are considered suitable for qualified presumption of safety (QPS) (EFSA2007b). In addition, these products have consumed for many years especially in rural areas of the country.

Fig. 2. Mashk

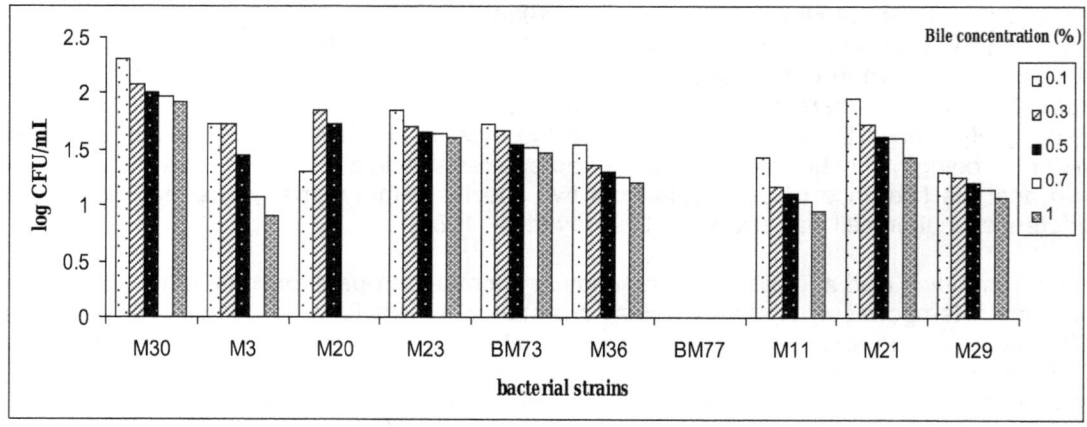

Fig. 3. Bile tolerance of selected LAB isolates as a control (time 0)

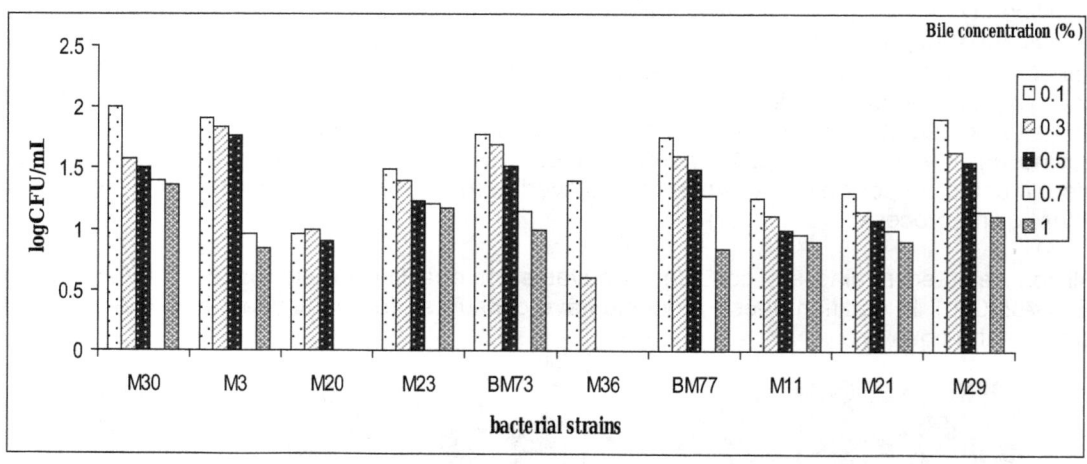

Fig. 4. Bile tolerance of selected LAB isolates after 4h

5. CONCLUSION

According to our knowledge, this is first report that states the presence of LAB strains in raw ewe milk, yoghurt and traditional sour buttermilk in Iran. This study proved the presence of viable non-starter LAB micro flora in these products. Among selected isolates, we found *Pediococcus acidilactici* in both milk and buttermilk, which indicates that these isolates were able to tolerate high processing temperatures and survive in final product (buttermilk). Furthermore, this strain might be added to product after boiling which comes from Mashk. The antagonistic activity possessed by these isolates might be used for the control of unwanted pathogens mainly in dairy products. Overall results of this research suggests that the selected LAB strains isolated might be appear to possess probiotic potential, and hence could be exploited further for their use in fermented dairy products

COMPETING INTERESTS

Authors have declared that no competing interests exist.

REFERENCES

Aslim, B., Yukesekdag, Z.N., Sarikaya, E., Beyatli, Y. (2005). Determination of the bacteriocin-like substances produced by some lactic acid bacteria isolated from Turkish dairy products. LWT- Food Science and Technology, 38, 691-694.

Badis, A., Guetarni, D., Moussa Boudjema, B., Henni, D.E., Kihal, M. (2004). Identification and technological properties of lactic acid bacteria isolated from raw goat milk of four Algerian races. Food Microbiology, 21, 579-588.

Bertazzoni Minelli, E., Benini, A., Marzotto, M., Sbarbati, A., Rauzzenente, O., Ferrario, R., Hendriks, H., Dellaglio, F. (2004). Assessment of novel *Lactobacillus casei* strains for the production of functional dairy foods. International Dairy Journal, 14, 723-736.

Christiansen, P., Nielsen, E.W.K., Vogensen, F., Bogren, C.H., Ardo, Y. (2006). Heat resistance of *Lactobacillus paracasei* isolated from semi-hard cheese made of pasteurized milk. International Dairy Journal, 16, 1196-1204.

Chou, L.S., Weimer, B. (1998). Isolation and Characterization of Acid- and Bile-Tolerant isolates from Strains of *Lactobacillus acidophilus*. Journal Dairy Science, 82(1), 23-31.

Conway, P.L., Gorbach, S.L., Goldin, B.R. (1987). Survaval of lactic acid bacteria in the human stomach and adhesion to intestinal cells. Journal of Dairy Science, 70 (1), 1-12.

De Angelis, M., Di Cagno, R., Huet, C., Crecchio, C.F., Fox, P., Gobbetti, M. (2003). Heat shock response in *Lactobacillus plantarum*. Applied and Environmental Microbiology, 1336-1346.

EFSA. (2007b). Opinion of the scientific committee on introduction of a qualified presumption of safety (QPS) approach for assessment of selected microorganisms referred to EFSA. EFSA Journal, 587,1e16.

EFSA. (2009a). Scientific opinion on the substantiation of health claims related to non-characterized microorganisms pursuant to article 13(1) of regulation (EC) no. 1924/20061.   Sci.   Opin.   Panel   Diet.   Prod.Nutr.Allerg. http://www.efsa.europa.eu:80/cs/BlobServer/Scientific_Opinion/nda_op_ej1247_art13 (1)_non_characterised_microorganisms_related_claims_en, 0.pdf? Ssbinary=true.

Erkkila, S., Petaja, E. (2000). Screening of commercial meat starter cultures at low pH and in the presence of bile salts for potential probiotic use. Meat Science, 55, 297-300.

Feldsine, P., Abeyta, C., Andrews, W.H. (2002). AOAC International Methods Committee. AOAC International methods committee guidelines for validation of qualitative and quantitative food microbiological official methods of analysis. J AOAC Int. 2002 Sep-Oct; 85(5), 1187-200.

Gilliland, S.E., Staley, T.E., Bush, L.J. (1984). Importance of bile tolerance of *Lactobacillus acidophilus* used as a dietary adjunct. Journal of Dairy Science, 67(12), 3045-3051.

Gurira, O.Z., Buys, E.M. (2005). Characterization and antimicrobial activity of *Pediococcus* species isolated from South African farm-style cheese. Food Microbiology, 22, 159-168.

Karimi Torshizi, M.A., Rahimi, S.H., Mojgani, N., Esmaelkhanian, S., Grimes, J.L. (2008). Screening of indigenous strains of lactic acid bacteria for development of a probiotic for poultry. Asian-Australasian Journal of Animal Science, 1011-2367.

Liong, M.T.,Shah, N.P. (2005). Acid and bile tolerance and the cholesterol removal ability of *Bifidobacteria* strains. Bioscience Microflora, 24(1), 1-10.

Mathara, J.M., Schillinger, U., Kutima, P.M., Mbugua, S., Guigas, C., Franz, C., Holzapfel, W.H. (2008). Functional properties of *Lactobacillus platarum* strains isolated from Maasai traditional fermented milk produced in Kenya. Current Microbiology, 56, 315-321.

Marth, E., Steele, J. (2001). Applied dairy Microbiology. Second edition. Marcel Dekker Inc.

Mobarez, A.M., Hosseini Doust, R., Sattari, M., Mantheghi, N. (2008). Antimicrobial effects of bacteriocin like substance produced by *L. acidophilus* from traditional yoghurt on *P. aeruginosa* and *S. aureus*. Journal of Biological Science, 8, 221-224.

Mohamadi, M., Ezzatpanah, H., MahdaviAdeli, H.R., Mohamadifar, M.A. (2012). The effect of lactation period on chemical composition and physical properties of traditional Iranian buttermilk. Food Technology and Nutrition, 9, 37-44.

De Vos, P.M., Garrity, G., Jones, D.R., Krieg, N., Ludwig, W.A., Rainey, F., Schleifer,K.H., B.Whitman, W.(2009). Bergey's Manual of Systematic Bacteriology. Second edition, vol. 3, Springer.

Tajabadi Ebrahimi, M.C., Ouwehand, A.A.Hejazi, M., Jafari, P. (2011). Traditional Iranian dairy products: A source of potential probiotic *Lactobacilli*. African Journal of Microbiology Research, 15(1), 20-27.

Tamime, A.Y., Robinson, R.K. (2007). Tamime and Robinsons yoghurt science and technology. Third edition. Woodhead publishing limited.

Yukeskdag, Z.N., Beyatli, Y., Aslim, B. (2004). Determination of some characteristics coccoid forms of lactic acid bacteria isolated from Turkish kefirs with natural probiotic. LWT-Food Science and Technology, 37, 663-667.

8

Effect of Gamma Irradiation on Antibacterial Properties of Sea Crab Shell Chitosan

F.C.K. Ocloo[1*], A. Adu-Gyamfi[1], E. A. Quarcoo[1], Y. Serfor-Armah[2], D. K. Asare[1] and C. Owulah[1]

[1]Biotechnology and Nuclear Agriculture Research Institute, Ghana Atomic Energy Commission, P.O. Box LG 80, Legon, Ghana.
[2]National Nuclear Research Institute, Ghana Atomic Energy Commission, P.O. Box LG 80, Legon, Ghana.

Authors' contributions

This work was carried out in collaboration between all authors. FCKO designed the study, performed the statistical analysis, wrote part of the protocol, and wrote part of the first draft of the manuscript. AAG also wrote part of the protocol and part of the first draft. EAQ helped in the irradiation of the chitosan solution. YSA, DKA and CO managed the analyses of the study. All authors read and approved the final manuscript.

ABSTRACT

Effect of gamma irradiation on antibacterial activities of chitosan is described. Chitosan was prepared from crab shells via demineralization, deproteinization, decoloration and deacetylation. Chitosan solutions (2%) were prepared in 1% acetic acid and irradiated at 0, 5, 15 and 25 kGy. The degree of deacetylation and viscosity-average molecular weight of the chitosan were determined. Susceptibility tests of *E. coli* and *S. parathyphi* against the chitosan were determined. *E. coli* was more susceptible to lower concentrations of chitosan solutions. Irradiated chitosan in solutions exerted a slightly faster inhibition on both *E. coli and S. parathyphi* than the unirradiated chitosan solution, but there was no difference observed between irradiated and unirradiated chitosan in solutions after 48 hours of incubation. The degree of susceptibility of both *E. coli* and *S. parathyphi* to irradiated chitosan in solutions was not significantly affected by the irradiation dose.

*Corresponding author: Email: fidelis_ocloo@yahoo.com, ocloofid@hotmail.com

Keywords: Antibacterial activity; susceptibility; inhibition; molecular weight; chitosan; irradiation.

1. INTRODUCTION

Chitosan, an amino polysaccharide, has received much attention as a functional biopolymer for many diverse applications in food, pharmaceutical and cosmetics (Kumar, 2000; Shahidi et al., 1999). In many of these applications, specific molecular weights (Mw) of polysaccharides are required. Chitosan with an average Mw in the range of 5–10 kDa possesses strong bactericidal and superior biological activities (Kittur et al., 2003). Chitosan of 20 kDa prevents progression of diabetes mellitus and exhibits higher affinity for lipopolysaccharides compared to 140 kDa chitosan (Kondo et al., 2000). Chitooligomers have special antimicrobial activity (Begona and Ruth, 1997; Zheng and Zhu, 2003) and antitumour activity (Qin et al., 2002).

Recently, the antimicrobial and antioxidant activities of chitosan and its derivatives have attracted attention. It is of great interest to degrade chitosan into low molecular weight fragments under appropriate conditions, as these low molecular weight chitosans possess useful biological activities. The antibacterial effects of chitosan and chitosan oligomers are reported to be dependent on the molecular weight, degree of deacetylation (DD), and the type of bacterium (Uchida et al., 1989; Jeon et al., 2001; No et al., 2002; Tsai et al., 2002). The mechanism of antimicrobial activity of chitosan and derivatives is still yet to be elucidated. However, chitosan molecules are reported to be stacked over the microbial cell surface, blocking the nutrients (Shon, 2001) or bind to DNA as such inhibiting transcription or permeability of the microbial cell wall (Tharanathan and Kittur, 2003). Kumar et al. (2007) reported of higher bactericidal activity against *Bacillus cereus* and *Escherichia coli* for homogeneous low molecular weight chitosans (LMWC) of molecular weight 9.5–8.5 kDa, obtained by pronase catalyzed non-specific depolymerization (at pH 3.5, 37°C) than native chitosan.

Low-molecular weight chitosan can be prepared by chemical, radiation, or enzymatic degradation of the high-molecular weight polymer. Radiation can provide a useful tool for degradation of different polymers. In the reaction, no other chemical reagents are introduced and there is not a need to control the temperature, environment or additives (Feng et al., 2008). Specifically, radiation can induce reactions such as chain scissions of the 1-4 glycosidic bonds which cause a reduction in molecular weight of the polymer and negligible cross-linking (Lim and Tung, 1997). Recent work by Feng et al. (2008) proved that lowering the molecular weight of chitosan increased antioxidant activity. However, research is necessary to determine the antimicrobial activity of chitosan as the molecular weight is reduced. The objective of this study was to assess the effect of gamma irradiation on the antibacterial activity of chitosan.

2. MATERIALS AND METHODS

2.1 Materials

2.1.1 Sample collection and preparation

Crab shells were purchased from fishermen in Accra, Ghana. The shells were washed and then dried in the oven at 60°C overnight. The dried shells were ground in a moulinex blender, sieved to a particle size of 90 μm and then packaged in polyethylene bag for storage at ambient temperature until used.

2.1.2 Reagents and media

All reagents used in the study were from Sigma-Aldrich Chemie GmbH (Taufkirchen, Germany). These reagents were used without any further purification. Microbiological media were from Oxoid, United Kingdom.

2.2 Isolation/Production of Chitosan

Chitosan was produced from crab shells using the methodology of No et al. (1989) with some modification.

2.2.1 Demineralization

Crab shells were demineralized with 1N HCl for 30 minutes at ambient temperature with a solid to solvent ratio of 1: 15 (w/v) (No et al., 1989) with constant stirring and then filtered under vacuum. The retentate was washed for 30 minutes with tap water and oven-dried.

2.2.2. Deproteinization

Demineralized shells were deproteinized with 3.5 % (w/w) NaOH solution for 2 hours at 65^0C with constant stirring at a solid to solvent ratio of 1: 10 (w/v) (No et al., 1989). The sample was filtered under vacuum, and the retentate washed with tap water for 30 minutes and oven-dried.

2.2.3 Decoloration

Demineralized and deproteinized Crab shells (crab chitin) were decolorized with acetone (1:10) for 10 minutes and dried for 2 hours at ambient temperature, followed by bleaching with 0.315 % (v/v) sodium hypochloride (NaOCl) solution (containing 5.25% available Chloride) for 5 minutes at ambient temperature with a solid to solvent ratio of 1: 10 (w/v), based on dry shell (No et al., 1989). Samples were then washed with tap water and dried under vacuum for 3 hours until the powder was crispy.

2.2.4 Deacetylation

Crab chitin was refluxed for 6 hours at 100°C using 50% concentrated sodium hydroxide solution (NaOH) with a solid to solvent ratio of 1: 15 (w/v). The resulting chitosan was washed to neutrality with tap water, rinsed with hot distilled water (90°C), filtered, and dried at 60°C for 24 hours in the oven.

2.3 Radiation

Samples of chitosan in solutions were irradiated at the Gamma Irradiation Facility at the Radiation Technology Centre (RTC) of the Ghana Atomic Energy Commission using a Co-60 source. Samples of 2g each were dissolved in 1% (v/v) acetic acid solution (100 mL), packaged in plastic container and samples irradiated at 0, 5, 15 and 25 kGy at a dose rate of 1.9846 kGy/hour.

2.4 Analysis

2.4.1 Degree of deacetylation

Degree of deacetylation was determined for unirradiated chitosan. Film prepared from the sample was used to study the degree of deacetylation (DD). The film was prepared by casting 1.0% w/v chitosan in 1% acetic acid solution, followed by drying in a vacuum air for 12 hr. The film was then deprotonated by washing 3 times with methanol and kept in a desiccator for 12hr, then placed in sealed plate before scanning. The spectra of the chitosan was obtained using a FTIR (Fourier Transform Infrared Spectroscopy) (FTIR-8400S CE, Shimadzu Corporation, Japan) with a frequency range of $4000 - 400$ cm^{-1}. The degree of deacetylation of the chitosan was calculated using the baseline developed by Sabnis and Block (1997):

$$DD = 97.67 - [26.486 \times (A_{1655} / A_{3450})]$$

where A_{1655} and A_{3450} are the absorbance at 1655 cm^{-1} of the amide-I band (a measure of the N-acetyl group content) and 3450 cm^{-1} of the hydroxyl band as an internal standard to correct for film thickness.

2.4.2 Viscosity-average molecular weight (Mv) determination

Five chitosan concentrations of 0.20, 0.10, 0.05, 0.025 and 0.0125% (w/v) were prepared in 1% acetic acid for each of the irradiated chitosan samples (0, 5, 15 and 25 kGy). The relative viscosity measurement was performed by using an Ubbelohde capillary viscometer (size 1, Poulten, Selfe & Lee Ltd, England) at $25\pm1°C$. The intrinsic viscosity is defined as:

$$[\eta] = (\eta_{red})c \rightarrow 0$$

Which is the value of the reduced viscosity (η_{red}) at zero concentration obtained from the linear plot of intrinsic viscosity against concentration. The viscosity molecular weight was calculated based on Mark Houwink equation (Chen and Hwa, 1996):

$$[\eta] = KM^a \text{ or } \log[\eta] = \log K + a \log M$$

where, $[\eta]$ is intrinsic viscosity, M is viscosity average molecular weight (in Dalton), K and a are empirical volumetric constants given by 8.93×10^{-4} cm/g and a = 0.71 respectively.

2.4.3 Antibacterial property of chitosan samples

2.4.3.1 Bacterial culture

Stock culture of *Escherichia coli* (*E. coli*) and *Salmonella parathyphi* (*S. parathyphi*) were reactivated on Eosin Methylene Blue Agar and Xylose Lysine Deoxycholate Agar, respectively, to obtain 24 hr cultures. An inoculum concentration between $10^7 - 10^9$ cfu/ml was made by inoculating colonies into sterile trypticase soya broth (TSB, Oxoid) and used for susceptibility test.

2.4.3.2 Susceptibility test

Susceptibility test was done by the tube dilution method described by Sugumar et al. (2010) with some modification. Stock solutions were 2% chitosan in solutions irradiated at 0, 5, 15 and 25 kGy. Varying concentrations of 0.02, 0.04, 0.06 and 0.2% were prepared for each dose treatment and the final pH adjusted to 6.8. One millilitre (1 ml) of the inocula was added to 9 ml of each chitosan solution prepared and incubated at 37°C. One millilitre (1 ml) of the incubated mixture was taken into 9 ml of TSB at intervals of 0, 6, 12, 24 and 48 hrs for pour plating using Eosin Methylene Blue Agar (Oxoid, UK) for *E. Coli* and Xylose Lysine Deoxycholate Agar (Oxoid, UK) for *S. parathyphi*. Plates were incubated at 37°C and counts were made between 18 – 24 hrs. For each chitosan solution, triplicate plating was undertaken.

2.5 Statistical Analysis

ANOVA was performed on the average-molecular weight data using MINITAB 14 (Minitab Inc., USA). The level of significance used was p<0.05 at 95% Confidence Intervals. Microsoft excel 2000 was used for graphical representation.

3. RESULTS AND DISCUSSION

3.1 Degree of Deacetylation

The degree of deacetylation as determined by FTIR was 80% (result not shown), which is comparable to values of 56 – 96% reported by No and Meyers (1995). Values ranging from 70 – 95% have also been reported (Canella and Garcia, 2001; Fernadez-Kim, 2004; Emi-Reynolds et al., 2007). Ocloo et al. (2011) have equally reported values of degree of deacetylation to be 76% for shrimp chitosan and 82% for commercial crab shell chitosan.

3.2 Viscosity-Average Molecular Weight (Mv) Determination

The molecular weight (M_v) of chitosan in solutions decreased significantly (p<0.05) with irradiation dose (Fig. 1) as a result of degradation. A sharp decrease was observed when irradiation dose was increased from zero (0) to 5 kGy. Similar findings were reported by Pasaphan et al. (2010).

3.3 Antibacterial Property of Chitosan Samples

The 1% Acetic acid without chitosan had no effect on the populations of both *E. coli* and *S. parathyphi* after 48 hours incubation (Tables 1 and 2). Also, unirradiated chitosan solutions of 0.02, 0.04 and 0.06% concentrations reduced the population of *E. coli* by 1 to 2 log units after 24 to 48 hours incubation (Table 1). However, the 0.2% concentration of unirradiated chitosan solution reduced the population of *E. coli* by 2 to 6 log units after 6 to 48 hours (Table 1). Chitosan solutions of 0.02, 0.04 and 0.06% concentrations from irradiated samples (5, 15 and 25 kGy) reduced populations of *E. coli* by up to 3 log units after 12 to 48 hours but the solution of 0.2% concentration greatly reduced the population of *E. coli* by 2 to 6 log units.

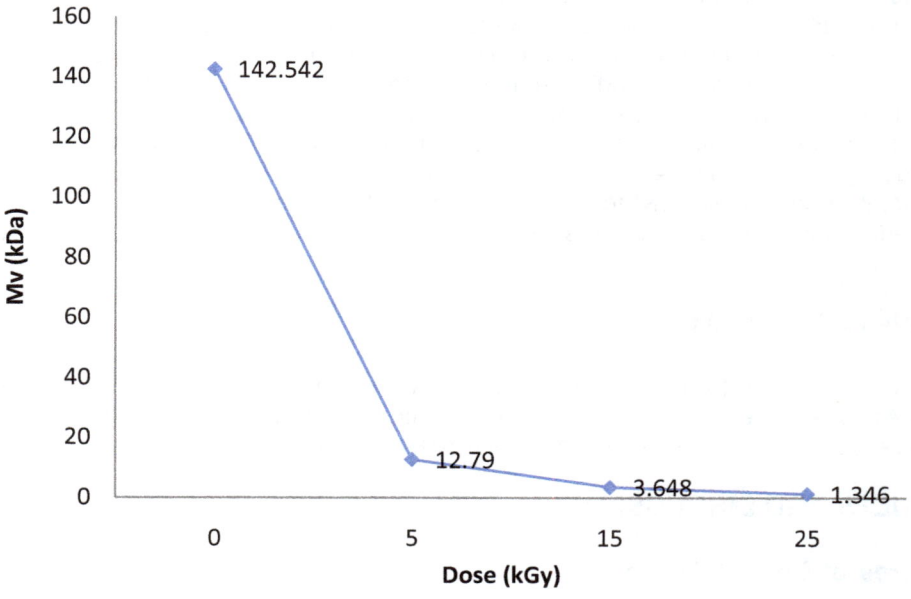

Fig 1: Effect of irradiation on viscosity-average molecular weight (Mv) of chitosan solutions

Table 1. Susceptibility of *E. coli* to irradiated chitosan solution at different concentrations

Dose (kGy)	Concentration of Chitosan (%)	0 hr	6 hrs	12 hrs	24 hrs	48 hrs
0	Control (1 % acetic acid)	$> 10^6$	$> 10^6$	$> 10^6$	$> 10^6$	$> 10^6$
	0.02	$> 10^6$	$> 10^6$	3.0×10^5	7.3×10^4	4.2×10^4
	0.04	$> 10^6$	$> 10^6$	$> 10^6$	5.2×10^6	2.3×10^5
	0.06	$> 10^6$	$> 10^6$	$> 10^6$	$> 10^6$	3.1×10^4
	0.20	$> 10^6$	1.7×10^4	1.3×10^4	0	0
5	Control (1 % acetic acid)	$> 10^6$	$> 10^6$	$> 10^6$	$> 10^6$	$> 10^6$
	0.02	$> 10^6$	$> 10^6$	7.5×10^5	8.3×10^4	6.0×10^4
	0.04	$> 10^6$	$> 10^6$	4.8×10^6	8.2×10^5	4.3×10^3
	0.06	$> 10^6$	4.0×10^5	1.8×10^4	3.4×10^4	6.7×10^3
	0.20	$> 10^6$	4.9×10^5	1.2×10^4	0	0
15	Control (1 % acetic acid)	$> 10^6$	$> 10^6$	$> 10^6$	$> 10^6$	$> 10^6$
	0.02	$> 10^6$	$> 10^6$	$> 10^6$	$> 10^6$	6.3×10^5
	0.04	$> 10^6$	$> 10^6$	3.0×10^5	4.9×10^5	3.8×10^4
	0.06	$> 10^6$	6.2×10^4	1.1×10^4	8.7×10^4	1.9×10^3
	0.20	$> 10^6$	1.5×10^5	3.0×10^1	0	0
25	Control (1 % acetic acid)	$> 10^6$	$> 10^6$	$> 10^6$	$> 10^6$	$> 10^6$
	0.02	$> 10^6$	$> 10^6$	1.2×10^6	$> 10^6$	8.1×10^4
	0.04	$> 10^6$	$> 10^6$	1.9×10^4	3.5×10^3	7.6×10^3
	0.06	$> 10^6$	4.0×10^6	1.3×10^5	1.5×10^4	8.3×10^3
	0.20	$> 10^6$	7.5×10^6	4.0×10^4	5.4×10^2	0

Each value is the average of 3 counts

On the other hand, unirradiated chitosan solutions of 0.02 and 0.04% concentrations had no effect on the population of *Salmonella parathyphi* after 48 hours incubation (Table 2). While the 0.06% concentration of unirradiated chitosan solutions gradually reduced population of *S. parathyphi* by 3 log units from 12 to 48 hours, the 0.2% concentration of unirradiated chitosan solution also reduced the population by 5 to 6 log units. Solutions of 0.02% concentration from irradiated chitosan (5, 15 and 25 kGy) had no inhibitory effect on the populations of *S. parathyphi* after 48 hours of incubation; however 0.04% concentration of chitosan solution irradiated samples (15 and 25 kGy) slightly reduced populations of *S. parathyphi* 1 to 2 log units after 24 to 48 hour. Solutions of 0.06% concentration from irradiated chitosan (5, 15 and 25 kGy) reduced populations of *S. parathyphi* by 1 to 3 log units after 24 to 48 hours. In the case of 0.2% concentration from irradiated chitosan in solution at 25 kGy, the population of *S. parathyphi* was reduced greatly by 1 to 6 log units from 6 to 48 hours.

Generally, all the irradiated chitosan solutions exerted a slightly faster inhibition (within 12 hours) on both *E. coli and S. parathyphi* than the unirradiated chitosan solution. However, there was no observable difference between the irradiated and unirradiated chitosan solutions after 48 hours of incubation.

The results of this study have shown that different concentrations of unirradiated and irradiated chitosan in solutions had varying degrees of inhibition against *E. coli and S. parathyphi*. However, 1% Acetic acid solution did not exert any noticeable effect on the population of the test isolates. These observations have demonstrated the antimicrobial activity of chitosan against *E. coli and S. parathyphi.*, confirming results reported by Chen et al. (1998), Rhoades and Roller (2000), Roller and Covill (2000) and Tsai et al. (2000).

The study has also shown that the degree of microbial inhibition of chitosan was dependent on its concentrations in solutions. This observation supports the findings of Seo et al. (2008). The study further showed that the degree of inhibition of the various chitosan solutions is dependent on the duration of incubation.

Despite the fact that there are several intrinsic and extrinsic factors that affect the antimicrobial activity of chitosan, the viscosity-average molecular weight has been identified as vital factor. Various studies have confirmed that irradiation of chitosan reduces the viscosity-average molecular weight (Yoksan et al., 2004; Feng et al., 2008) and that chitosan with lower molecular weight (of less than 10kDa) have greater antimicrobial activity than native chitosans (Uchida, 1989). In this study also, an irradiation dose of 5 kGy significantly decreased molecular weight of chitosan but this did not proportionately enhance the antimicrobial activity. Although both *E. coli* and *S. parathyphi* were found be susceptible to irradiated and unirradiated chitosan solutions, the rate of inhibition of these test isolates was marginally increased by irradiation.

Generally, the study suggests that *E. coli* was more susceptible to lower concentrations of chitosan compared to *S. parathyphi* and that the degree of susceptibility of *E. coli* and *S parathyphi* to irradiated chitosan in solutions was not significantly affected by the irradiation dose.

Table 2. Susceptibility of *Salmonella paratyphi* to irradiated chitosan solution at different concentrations

Dose (kGy)	Concentration of Chitosan (%)	0 hr	6 hrs	12 hrs	24 hrs	48 hrs
0	Control (1% acetic acid)	$> 10^6$	$> 10^6$	$> 10^6$	$> 10^6$	$> 10^6$
	0.02	$> 10^6$	$> 10^6$	$> 10^6$	$> 10^6$	$> 10^6$
	0.04	$> 10^6$	$> 10^6$	$> 10^6$	$> 10^6$	$> 10^6$
	0.06	$> 10^6$	$> 10^6$	4.4×10^4	1.9×10^4	6.4×10^3
	0.20	$> 10^6$	$> 10^6$	6.3×10^5	2.0×10^1	0
5	Control (1 % acetic acid)	$> 10^6$	$> 10^6$	$> 10^6$	$> 10^6$	$> 10^6$
	0.02	$> 10^6$	$> 10^6$	4.6×10^6	$> 10^6$	$> 10^6$
	0.04	$> 10^6$	$> 10^6$	$> 10^6$	$> 10^6$	$> 10^6$
	0.06	$> 10^6$	4.0×10^6	8.6×10^5	1.3×10^5	3.6×10^3
	0.20	$> 10^6$	6.4×10^5	2.5×10^2	1.0×10^1	0
15	Control (1 % acetic acid)	$> 10^6$	$> 10^6$	$> 10^6$	$> 10^6$	$> 10^6$
	0.02	$> 10^6$	$> 10^6$	$> 10^6$	$> 10^6$	$> 10^6$
	0.04	$> 10^6$	$> 10^6$	$> 10^6$	4.7×10^5	7.3×10^4
	0.06	$> 10^6$	$> 10^6$	4.1×10^6	6.2×10^3	1.8×10^3
	0.20	$> 10^6$	2.3×10^5	4.5×10^2	4.0×10^1	0
25	Control (1 % acetic acid)	$> 10^6$	$> 10^6$	$> 10^6$	$> 10^6$	$> 10^6$
	0.02	$> 10^6$	$> 10^6$	$> 10^6$	$> 10^6$	$> 10^6$
	0.04	$> 10^6$	$> 10^6$	7.6×10^6	4.1×10^6	8.6×10^5
	0.06	$> 10^6$	$> 10^6$	2.7×10^4	5.0×10^3	2.2×10^3
	0.20	$> 10^6$	3.1×10^5	3.1×10^4	4.2×10^2	0

Each value is the average of 3 counts

4. CONCLUSION

Microorganisms such as *E. coli and S. parathyphi* are susceptible to chitosan solution and this could be explored to improve food safety and stability. *E. coli* was more susceptible to lower concentrations of chitosan solutions and the degree of inhibition of both *E. coli* and *S parathyphi* was marginally increased by irradiation. Irradiation decreased molecular weight of chitosan but the degree of susceptibility of *E. coli* and *S parathyphi* to irradiated chitosan solutions was not significantly affected by the irradiation dose (0, 5, 15 and 25 kGy).

ACKNOWLEDGEMENT

This project was executed under the IAEA funded project CRP-RC-14730/R. We wish to express our profound gratitude to IAEA for their financial support. Also to all technicians, technologists, trainees and national service personnel of Radiation Technology Centre (RTC) especially, Mr. Daniel Larbi, Shadrach, Sylvester, Riverson and Ernestina. as well as staff of Gamma Irradiation Facility of RTC.

COMPETING INTERESTS

Authors have declared that no competing interests exist.

REFERENCES

Begona, C.G., Ruth, D. (1997). Evaluation of the biological properties soluble chitosan and chitosan microspheres. International Journal of Pharmaceutics, 148, 231–240.

Canella, K.M.N., Garcia, R.B. (2001). Characterization of chitosan by gel permeation chromatography-influence of preparation method and solvent.Quim. Nova, 24(1), 13-17.

Chen, C.S., Liau, W.Y., Tsai, G.J. (1998). Antibacterial effects of N-sulfonated and N-sulfobenzoyl chitosan and applications to oyster preservation. J. Food Prot. 61, 1124–1128.

Chen, R.H., Hwa, H.D. (1996). Effect of molecular weight of chitosan with the same degree of deacetylation on the thermal, mechanical and permeability properties of the prepared membrane. Carbohydr. Polym., 29, 353-358.

Emi-Reynolds, G., Zaki, S., Banini, G.K., Dogbe, S.A., Ofori-Appiah, M.A. (2007). Radiation processing and characterization of chitin and chitosan extracted from crab shells. Journal of the Ghana Science Association, 9(2), 18 – 24.

Feng, T., Li, J.D.Y., Hu, Y.,Kennedy, F.K. (2008). Enhancement of antioxidant activity of chitosan by irradiation.Carbohydrate Polymers, 73(1), 126-132.

Fernadez-Kim, S.O. (2004). Physico-chemical and functional properties of crawfish chitosan as affected by different processing protocols. M.Sc. Thesis, Louisiana State University, USA, 76.

Jeon, Y.J., Park, P.J., Kim, S.K. (2001). Antimicrobial effect of chitooligosaccharides produced by bioreactor. Carbohydr Polym, 44, 71–6.

Kittur, F.S., Vishu Kumar, A.B.,Tharanathan, R.N. (2003). Low molecular weight chitosans – preparation by depolymerization with Aspergillus niger pectinase and characterization. Carbohydrate Research, 338, 1283–1290.

Kondo, Y., Nakatani, A., Hayashi, K., Ito, M. (2000). Low molecular weight chitosan prevents the progression of low dose streptozotocininduced slowly progressive diabetes mellitus in mice. Biological & Pharmaceutical Bulletin, 23, 1458–1464.

Kumar, M.N.V.R. (2000). A review of chitin and chitosan applications. Reactive & Functional Polymers, 46, 1–27.

Kumar, A.B.V., Varadaraj, M.C., Gowda, L.R.,Tharanathan, R.N. (2007). Low molecular weight chitosans—Preparation with the aid of pronase, characterization and their bactericidal activity towards Bacillus cereus and Escherichia coli. Biochimica et Biophysica Acta, 1770, 495–505.

Lim, L.T., Tung, M.A. (1997). Vapor pressure of allyl isothiocyanate and its transport in PVDC/PVC copolymer packaging film.Journal of Food Science, 62, 1061–1066.

No, H.K., Park, N.Y., Lee, S.H., Meyers, S.P. (2002). Antibacterial activity of chitosans and chitosan oligomers with different molecular weights. Int J Food Microbiol, 74, 65–72.

No, H.K., Meyers, S.P. (1995). Preparation and Characterization of Chitin and Chitosan-A Review. Journal of Aquatic Food Product Technology, 4(2), 27-52.

No, H.K., Meyers, S.P., Lee, K.S. (1989). Isolation and characterization of chitin from crawfish shell waste. Journal of Agricultural and Food Chemistry, 37(3), 575-579.

Ocloo, F.C.K., Quayson, E.T., Adu-Gyamfi, A., Quarcoo, E.A., Asare, D., Serfor-Armah, Y., Woode, B.K. (2011). Physicochemical and functional characteristics of radiation-processed shrimp chitosan. Radiation Physics and Chemistry, 80, 837–841.

Pasanphan, W., Rimdusit, P., Choofong, S., Piroonpan, T., Nilsuwankosit, S. (2010). Systematic fabrication of chitosan nanoparticle by gamma irradiation. Radiation Physics and Chemistry, 79, 1095–1102.

Qin, C.Q., Du, Y.M., Xiao, L., Li, Z.,Gao, X.H. (2002). Enzymic preparation of water-soluble chitosan and their antitumor activity. International Journal of Biological Macromolecules, 31(3), 111–117.

Rhoades, J., Roller, S. (2000). Antimicrobial actions of degraded and native chitosan against spoilage organisms in laboratory media and foods. Appl. Environ. Microbiol.,66, 80–86.

Roller, S., Covill, N. (2000). The antimicrobial properties of chitosan in mayonnaise and mayonnaise-based shrimp salads. J. Food Prot., 63, 202–209.

Sabnis, S., Block, L.H. (1997). Improved infrared spectroscopic method for the analysis of degree of N-deacetylation of chitosan. Polym Bull., 39, 67-71.

Seo, S., King, J.M., Prinyawiwatkul, W., Janes, M. (2008). Antibacterial activity of ozone-depolymerized crawfish chitosan.J Food Sci., 73(8), M 400-4.

Shahidi, F., Arachchi, J.K.V., Jeon, Y.J. (1999). Food applications of chitin and chitosans. Trends in Food Science & Technology, 10, 37–51.

Shon, D. H. (2001). Chitosan oligosaccharides for functional foods and microbial enrichment of chitosan oligosaccharides in soy-paste, Proceedings of the International workshop on Bioactive Natural Products, Tokyo, Japan, 56–66.

Sugumar, G., Ramesh, U., Selvan, A. (2010). Susceptibility of Crab Chitosan against Staphylococcus aureus. Bioresearch Bulletin, 1, 7-9.

Tharanathan, R.N., Kittur, F.S. (2003). Chitin—The undisputed biomolecule of great potential, Crit. Rev. Food Sci. Nutr. 43, 61–87.

Tsai, G.J., Wu, Z.Y., Su, W.H. (2000). Antibacterial activity of a chitooligosaccharide mixture prepared by cellulase digestion of shrimp chitosan and its application to milk preservation. J. Food Prot., 63, 747–752.

Tsai, G.J., Su, W.H., Chen, H.C., Pan, C.L. (2002). Antimicrobial activity of shrimp chitin and chitosan from different treatments and applications of fish preservation. Fish Sci, 68, 170–7.

Uchida, Y., Izume, M., Ohtakara, A. (1989). 'Preparation of chitosan oligomers with purified chitosanase and its application'. In: Skjåk-Bræk G et al., editors. Chitin and chitosan: sources, chemistry, biochemistry, physical properties and applications. London, UK: Elsevier Applied Science, 373–82.

Yoksan, R., Akashi, M., Miyata, M., Chirachanchai, S. (2004). Optimal γ-Ray dose and irradiation conditions for producing low-molecular-weight chitosan that retains its chemical structure.Radiation Research, 161(4), 471-480.

Zheng, L.Y.,Zhu, J.F. (2003). Study on antimicrobial activity of chitosan with different molecular weights. Carbohydrate Polymers, 54, 527–530.

9

Galacturonic Acid Content and Degree of Esterification of Pectin from Sweet Potato Starch Residue Detected Using ^{13}C CP/MAS Solid State NMR

Siti Nurdjanah[1*], James Hook[2], Jane Paton[3] and Janet Paterson[4]

[1]Department of Agriculture Product Technology, University of Lampung, Lampung 35145, Indonesia, and a former Ph.D Research Student at Food Science and Technology, School of Chemical Engineering, Faculty of Engineering The University of New South Wales, Sydney, NSW 2052 Australia.
[2]NMR Facility, The University of New South Wales, Sydney, NSW 2052 Australia.
[3]Faculty of Science, The University of New South Wales, Sydney, NSW 2052, Australia.
[4]Food Science and Technology, School of Chemical Engineering, Faculty of Engineering The University of New South Wales, Sydney, NSW 2052, Australia.

Authors' contributions

This work was carried out in collaboration between all authors. Author SN designed the study, performed the laboratory work, the statistical analysis, and wrote the first draft of the manuscript. Author JH was responsible for the acquisition and interpretation of the NMR spectra. Authors JP and Janet P performed critical reviews of the first and draft. All authors read approved the final manuscript.

ABSTRACT

Starch residue samples from two Australian sweet potato varieties (Beauregard and Northern Star) and two Indonesian sweet potato varieties (Bis192 and Bis183), and a commercial sample of sweet potato starch residue, were studied for their pectins. Pectins were extracted using 0.1M HCl, 0.05M NaOH, 0.1M HCl/0.75% SHMP and 0.05M NaOH/0.75% SHMP. Hydrolysis of residual starch in the cell wall of sweet potato using heat stable α-amylase and amyloglucosidase was employed prior to pectin extraction to eliminate starch contamination. Pectins were characterised for yield, galacturonic acid content (GA), and the degree of esterification (DE). Conventionally, pectin is characterized

Corresponding author: Email: nurdjanah_thp@unila.ac.id

by titration, photometry and HPLC. However these methods are cumbersome and time consuming. On the other hand, ^{13}C CP/MAS solid-state NMR, a non-destructive, efficient and direct method, has been found to be well-suited for these purposes since pectin has well-defined ^{13}C NMR spectra. Therefore ^{13}C CP/MAS solid state NMR was used for pectin determination. The pectin characteristics are dependent on variety and extraction process; however, the extraction methods gave variable results. Yields were between 7 and 30% of the cell wall. GA varied from 27 to 80% with the highest found in Bis192 extracted using NaOH/SHMP. DE varied between traceable and 57%. HCl extraction gave higher DE, while NaOH/SHMP caused demethylation. Overall, this study demonstrated that pectin from sweet potato starch residue is mainly low in methoxyl groups.

Keywords: Sweet potato; starch residue; pectin; ^{13}C CPMAS solid state NMR.

1. INTRODUCTION

Pectin has many important functions in plants. It contributes to the structural integrity and mechanical strength of the tissue by forming a hydrated cross-linked three-dimensional network [1,2]. Pectin also plays an important role in the physical and sensory properties of fresh fruit and vegetables (ripeness and texture) and contributes to their processing characteristics in canned products, purees, and juices [3]. Commercially, pectin has broad applications in both the food and pharmaceutical industries, where it acts as gelling and thickening agents [2,4], prevents the formation of cheesy milk layer in gelled milk dessert, and regulates the thickness and mouth-feel of fruit drink powder when the powder is dissolved in cold water [5]. In addition, pectin has proven to have beneficial effects on human health [6,7,8].

To date, citrus peel and apple pomace are the major commercial sources of pectin. Many attempts have been made to prepare pectin from other sources such as tropical fruits [9], sunflower heads[10], beet and potato pulp [11], soy hull [12], and duckweed[13]. However, pectins extracted from those materials have poor gelling ability characteristic as compared to apple and citrus pectin.

The physico-chemical properties of beet pectin has been reported to be influenced by extraction conditions [14,15]. Sugar beet pectin that is not yet utilized fully due to poor gelling ability has been reported to have effective emulsifying properties [16,17]. Recently, Byg et al. [18] reported that industrial potato waste contains appreciable amount of rhamnogalacturonan I (hairy region of pectin). This opens the possibility to investigate the potential use of other crop residue materials, such as sweet potato starch residue, as pectin sources.

Sweet potato (*Ipomoea batatas* (L.) Lam), a fairly drought-tolerant crop, is widely grown throughout the world, primarily in the tropics and subtropics. In Indonesia, although sweet potato production is not as high as that of China, the trend with respect to utilization of sweet potato is changing from domestic consumption to use in various commercial products.

The utilization of sweet potato within the industrial sector, has led to the production of considerable amounts of waste materials such as starch residues all year around. Sweet potato starch residue (non-starch polysaccharide) has an appreciable amount of pectin [7,19,20,21,22]. However, in contrast to pectin from other plant sources, sweet potato pectin has never been studied intensely. Therefore the purpose of this research was to elucidate some chemical properties of pectin extracted from some varieties of Indonesian and Australian sweet potatoes.

Pectin is embedded between the matrix of starch and cell wall and these have similar solubility in the extraction media examined [23]. Therefore removal of starch from the sweet potato starch residue was attempted before pectin extraction. In this study, the sweet potato starch residue was prepared in the laboratory from the Australian sweet potato varieties. These isolates and the sweet potato starch residue from Indonesian varieties were subjected to starch removal procedures. There are several ways of removing residual starch from the cell wall. However, in order to minimize pectin degradation during starch-free cell wall preparation, the enzymatic hydrolysis method was chosen

2. MATERIALS AND METHODS

2.1 Raw Materials

Beauregard and Northern Star varieties of sweet potato were grown and packed by Kidd Enterprise, Redland Bay Queensland, and ordered via Yep Lum and Co. Stand 281, C Block, Flemington Market, NSW, Australia. BIS 183 and BIS 192 were obtained from Balai Penelitian Kacang Kacangan dan Umbi Umbian (Tuber and Legume Research Center) The ministry of Agriculture, Malang East Java, Indonesia, Australia in the form of dried chips/slices. Sweet potato starch residue was obtained from a local sweet potato manufacturer at South Lampung district, Province of Lampung Indonesia. Brought to Food Science UNSW, Sydney in the form of dried powder. The varieties processed were commercially grown in the surrounding areas, varieties were not known exactly, and consisted of mixed local varieties.

2.2 Cell Wall Materials (CWM) Preparation

CWM was prepared from sweet potato starch residue according to the method of Noda et al. [20] with a slight modification, where the incubation time in boiling water was reduced from 20 min to 5 min. In addition, glucoamylase (synonym amyloglucosidase: exo-I,4-α-glucan glucanohydrolase, EC 3.2. 1. 3) was employed in the second digestion. Ground dried sweet potato starch residue (100 g) was suspended in distilled water (200 mL) and boiled for 5 minutes. The suspension was maintained at 80°C, and 0.5 mL of heat-stable α-amylase (Termamyl 120 type LS from Novo Nordisk Denmark) was added, and then incubated for 30 min to hydrolyse the residual starch. The enzyme activity was 120 KNU/g (KNU is Novo units α-amylase- that is the amount of enzyme that breaks down 5-26 g of starch per hour at Novo's standard method). The mixture was centrifuged at 3000 rpm for 10 min, supernatant was discarded and digestion of the residue was repeated with 0.5 mL glucoamylase (EC 3.2.1.3 from *Aspergillus niger*, SIGMA, 30-60 units per mg protein). One unit will liberate 1 mg of glucose from starch in 3 min at pH 4.5 at 55°C. The mixture was filtered using two layers of cheesecloth. The residue was washed with distilled water, methanol and acetone, successively, and air-dried (Fig. 1).

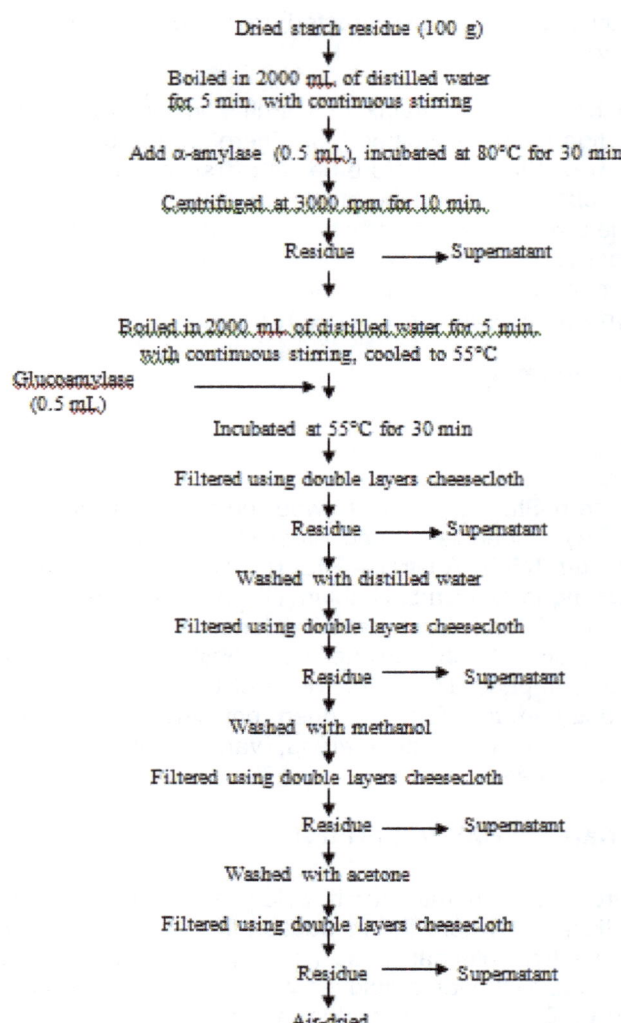

Fig. 1. Flow chart of cell wall preparation from starch residue

2.3 Pectin Extraction

Pectins were extracted from the cell wall by using solutions, namely 0.1 M hydrochloric acid, 0.05 M sodium hydroxide, 0.1 M HCL containing 0.75% sodium hexametaphosphate, and 0.05 M sodium hydroxide containing 0.75% sodium hexametaphosphate. These pectin extraction methods were slightly modified from Turquois et al. (1997) where the pectin was not extracted directly from the sweet potato pulp, instead, the pulp was previously freed from residual starch, using procedure described by Noda et al. (1994) with slight modification as described in Section 2.2. This non-starch residue refers to alcohol insoluble residue [24,25].

2.3.1 0.1 M HCl extraction

Samples (10 g) of dried ground cell wall materials were dispersed in 250 mL 0.1 M HCl. The dispersion was stirred and kept at 90°C for 1 hour. After incubation, the suspensions were centrifuged at 10°C for 15 min at 10000 rpm. The liquid fraction containing extracted pectin materials was neutralised with 32% NaOH (Laboratory UNILAB Reagent AJAX), then the same volume of 95% ethanol was added, the mixture was stirred for 5 minutes and then stored at 4°C for 12 hours. The mixture was then centrifuged at 10000 rpm for 15 min and the pectin residue washed with 70, 80, 90% ethanol, successively. Finally the extracted pectin was dried in a freeze dryer for 18 hours, ground and then stored in a desiccator (Fig. 2) prior to analysis for its galacturonic acid content, degree of esterification and starch content

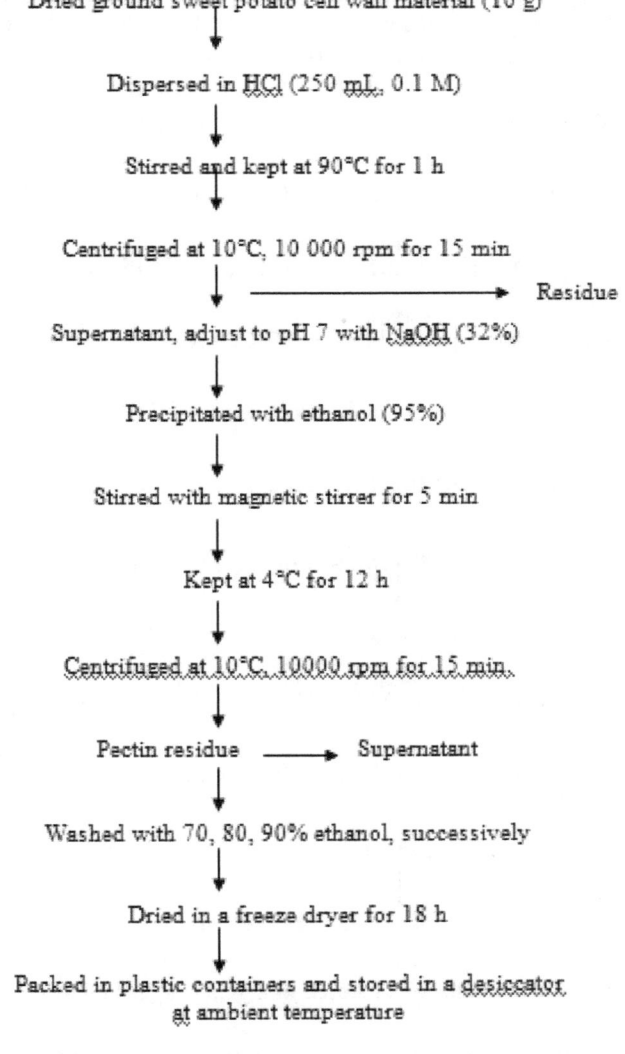

Fig. 2. Flow chart of pectin extraction using 0.1M HCl

2.3.2 Extraction using 0.1 M HCl and 0.75% sodium hexaxmetaphosphate

Pectin was extracted by using 0.1M HCl containing 0.75% sodium hexametaphosphate, using the procedure described in Section 2.3.1.

2.3.3 0.05 M NaOH extraction

Samples (10 g) of dried ground cell wall materials were dispersed in 250 mL 0.05M NaOH. The mixture was kept for 2 hours at 25°C, then centrifuged at 10°C, 10000 rpm for 15 minutes. The liquid fraction was neutralised with 5M HCl, and the same volume of 95% ethanol was added, stirred for 5 minutes then stored at 4°C for 12 hours. The mixture was centrifuged at 10°C, 10000 rpm for 15 min to separate the precipitated pectin from the ethanol solution. The precipitated pectin was washed successively with 70, 80, and 90% ethanol. The mixture was centrifuged at 10°C, 10000 rpm for 15 min, then dried in a freeze dryer for 18 hours. The dried pectin was ground, packed and stored in a desiccator until further analysis. A flow chart of this process is shown in Fig. 3.

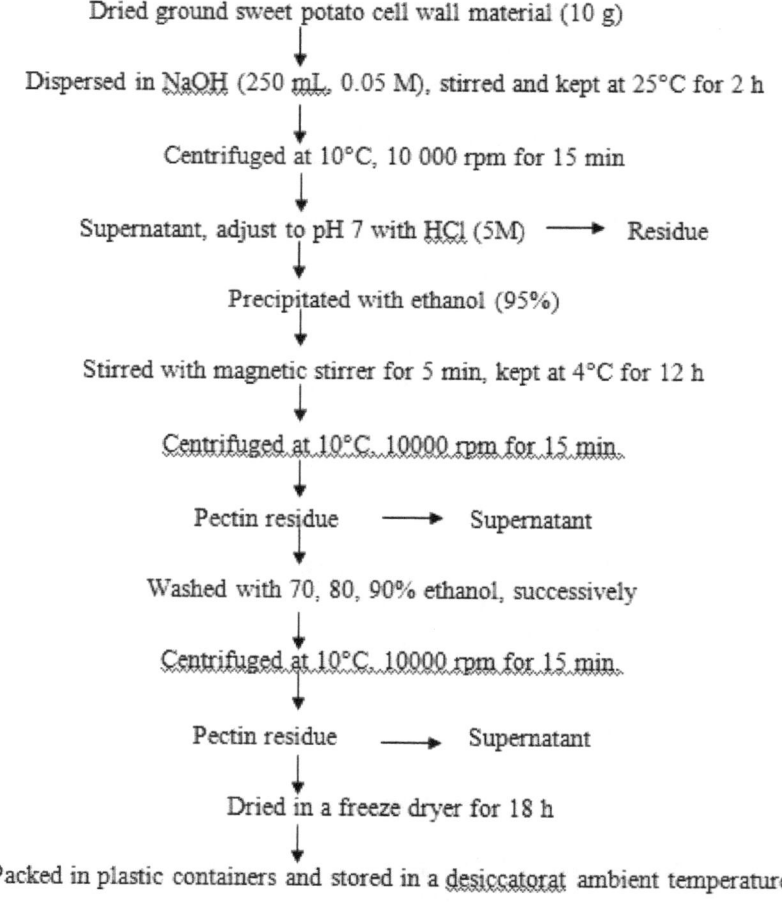

Dried ground sweet potato cell wall material (10 g)

Dispersed in NaOH (250 mL, 0.05 M), stirred and kept at 25°C for 2 h

Centrifuged at 10°C, 10 000 rpm for 15 min

Supernatant, adjust to pH 7 with HCl (5M) ⟶ Residue

Precipitated with ethanol (95%)

Stirred with magnetic stirrer for 5 min, kept at 4°C for 12 h

Centrifuged at 10°C, 10000 rpm for 15 min.

Pectin residue ⟶ Supernatant

Washed with 70, 80, 90% ethanol, successively

Centrifuged at 10°C, 10000 rpm for 15 min.

Pectin residue ⟶ Supernatant

Dried in a freeze dryer for 18 h

Packed in plastic containers and stored in a desiccator at ambient temperature

Fig. 3. Flow chart of pectin extraction using 0.05 M NaOH

2.3.4 0.05M NaOH and 0.75% sodium hexametaphosphate extraction

Pectin was extracted by using 0.05M NaOH containing 0.75% sodium hexametaphosphate, using the procedure described in Section 2.3.2.

2.4 Pectin Yield

Pectin yield was calculated as the ratio of dried pectin extracted to dried cell wall materials.

2.5 Galacturonic Acid and Degree of Esterification

The content of galacturonic acid and the degree of esterification were determined by using ^{13}C solid-state NMR method described as follows: ~300 mg of powdered samples were packed into partially-stabilsed zirconia rotors, sealed with fluted caps, then inserted into 7.5 mm CP/MAS Chemagnetics probe and spun at the magic angle (spinning at 54.7° between the direction of the static magnetic field and the rotor axis to reduce the chemical shift anisotropy/CSA pattern to its isotropic) at 4 kHz. Spectra were acquired using a Varian Inova 300 wide-bore solid state NMR spectrometer (Varian Associates, USA). Typical conditions were: observe frequency, 75.4 MHz, recycle delay 5 s, contact time 1 ms, sweep width of 50 kHz. The chemical shifts were adjusted using secondary referencing to hexamethylbenzene (HMB, methyl groups set to 17.3 ppm).

The ^{13}C CP/MAS spectra of non-methylated polygalacturonic acid and methylated citrus pectin (Sigma, Australia) were used as the basis for the interpretation pectin spectra. The signal at 172.7 ppm was assigned to the C-6 carbon of the COOH group and the intense signal at 101.9 ppm was assigned to the C-1 carbon. The peaks between 60 and 90 ppm are carbons of pyranoid rings (C-2,3,5), and the signal at 80.3 is from the C-4 carbon. The signal at 53.7 ppm is assigned to the methyl carbon of the methyl ester groups (COO$\underline{C}H_3$), and the resonances between 18 and 20 ppm are assigned to acetyl ester groups (OC$\underline{C}H_3$). The formulas recommended by Sullivan [26] and Sinitsya [27] were used for GA and DE calculation as follows: Galacturonic Acid (%) = (Area C-6/average area of C1-C5) X 100. Degree of Esterification (%)=(COO$\underline{C}H_3$/average area of C1-C5) X 100. The area of resonance of C-1 is converted to 100 and the areas of other resonances are relatively based on that of C-1. The calculation of peak area was performed using SOLARIS 2.7 software program.

2.6 Statistical Analysis

The experiment for cell wall material extraction was constructed as a complete randomized design with 3 replications, whereas the experiment for pectin extraction was constructed as a factorial in a complete randomised design with 3 replications. The first factor was sweet potato variety and the second factor was method of extraction. Analysis of variance (ANOVA) was used to analyse the data, and the comparison of means was carried out at the 5% significance level using the least significant different (LSD) test according to Steel and Torrie[28].

3. RESULTS AND DISCUSSION

3.1 Cell Wall Material of Sweet Potato

The content of cell wall material from the different sweet potato varieties varied from 35 to 52% of dry starch residue (Table 1). These results are higher than those obtained for different sweet potato varieties by Noda et al. [20] where the cell wall material was 33% of dried starch residue. This difference was attributed to method of starch extraction and different varieties. Noda et al. [20] extracted starch from sweet potato flesh by sieving, leading to a higher residual starch content in the residue and lower cell wall material content.

Table 1. Cell wall material content of sweet potato starch residue

Variety	Weight of dry starch residue (g)	Weight of cell wall material (g)	Percentage of cell wall material from dry starch residue*
Beauregard	100	35.3	35.3 a
Northern Star	100	40.6	40.6 b
Bis 192	100	41.6	41.6 b
Bis 183	100	43.3	43.3 b
Starch residue	100	52.1	52.1 c

Means within columns followed by the same letter are not significantly different (P>=0.05)

Beauregard sweet potato had the lowest cell wall material content which explains why this variety is famous for its texture for being less fibrous, soft and moist, with good eating quality (Kidd Enterprise, Queensland, Australia, Personal communication). The factory sample of sweet potato starch residue had the lowest starch content (33%, data not shown) but a higher cell wall material content than the starch residue sample prepared in the laboratory, which was also reported by Salvador et al. [21].

The separation of residual starch from the cell wall is very important because it gives rise to significant contamination during pectin extraction. In general, there are two different methods for starch hydrolysis: enzyme treatment, such as the use of α-amylase, and chemically, such as the use of dimethyl sulphoxide (DMSO). Extraction of starch with DMSO has disadvantages such as the loss of about 6% of the cell wall, mainly pectin [29], incomplete removal of starch (Noda *et al.* 1994), and safety problems because it is a skin, eye and respiratory irritant [30]. Therefore, a combination of heat stable α-amylase and glucoamylase was employed consecutively in this experiment.

Alpha-amylase (1,4-α-D-glucan glucanohydrolase) is an endo-enzyme that catalyses the hydrolysis of 1,4-α-D glucosidic bonds in a random fashion along the polysaccharide chains, whereas glucoamylase catalyses the hydrolysis of terminal 1,4-linkaged α-D-glucose residues successively from the non-reducing ends of malto-oligo and polysaccharides with release of β-D-glucose. Most forms of the enzyme can rapidly hydrolyse 1,6-α-D-glucosidic bonds when the next bond in the sequence is 1,4. α-Amylase alone was not sufficient to hydrolyse all the residual starch in the cell wall, as indicated by a positive reaction with I_2KI (Lugol's solution). Therefore, glucoamylase was employed for subsequent starch removal, mainly the branched polymer amylopectin.

3.1 Pectin yield

The yields of pectin extracted from cell wall materials of sweet potato using various conditions were between 7.2 and 29.3% of dry CWM, or between 0.3 and 1.2 % of sweet potato fresh weight. There are significant differences among varieties and treatments. Northern Star gave the highest yield, followed by Beauregard, Bis 192, Bis 183, and starch residue (Table 2).

Table 2. The yield of sweet potato pectin extracted using different conditions

Varieties	Extraction methods	Pectin yield (g)	Pectin yield (% of cell wall material)*
Beauregard	0.1M HCl	1.47	14.7 fg
	0.1M HCl cont. 0.75%SHMP	1.60	16.0 e
	0.05M NaOH	0.98	9.8 ij
	0.05M NaOH cont.0.75% SHMP	2.88	28.8 b
Northen Star	0.1M HCl	1.53	15.3 ef
	0.1N HCl cont. 0.75%SHMP	1.77	17.7 d
	0.05M NaOH	1.11	11.1 h
	0.05M NaOH cont.0.75% SHMP	3.00	30.0 a
Bis192	0.1M HCl	1.04	10.4 hi
	0.1M HCl cont. 0.75%SHMP	1.37	13.7 g
	0.05M NaOH	0.74	7.4 m
	0.05M NaOH cont.0.75% SHMP	2.78	27.8 b
Bis 183	0.1M HCl	0.93	9.3 jk
	0.1M HCl cont. 0.75%SHMP	1.11	11.1 h
	0.05M NaOH	0.86	8.6 kl
	0.05M NaOH cont.0.75% SHMP	2.86	28.6 b
Starch residue	0.1M HCl	0.89	8.9 jkl
	0.1M HCl cont. 0.75%SHMP	0.81	8.1 lm
	0.05M NaOH	0.72	7.2 m
	0.05M NaOH cont.0.75% SHMP	2.47	24.7 c

Means within columns followed by the same letter are not significantly different ($P = 0.05$).

The pectin yield from the industrial starch residue was considerably lower than that from laboratory-prepared starch residue possibly because there was a delay of one hour at around 35°C involved in the transportation of the sample from the starch-processing centre to the laboratory in the case of the former. Activity of endogenous pectinases during this delay may have led to degradation of some pectic substances.

Conditions of extraction also significantly affected pectin yields. In all varieties, the alkali/SHMP combination provided the most efficient procedure, followed by acid containing SHMP, and acid extraction. The alkali without SHMP gave the lowest yield. The low pectin yield in acid extraction was in contrast to other reports, which reported that generally, the highest pectin yields were obtained by hot acid extraction [31,32,33].

The effectiveness of alkali and chelating agents such as SHMP as extraction media for pectin involves two mechanisms, chelation of Ca^{2+} by SHMP and destruction of the alkali-labile linkages such as esters, some glycosidic linkages between methoxylated galacturonic residues and hydrogen bonds [34]. This indicates that pectins in sweet potato cell walls are mainly calcium-bound low methoxyl pectin that is not extractable with mild acid or alkaline [35].

3.3 Galacturonic Acid, Degree of Esterification and Acetylation of Pectin

The ^{13}C CP/MAS NMR spectra of polygalacturonic acid and citrus pectin (Sigma, Australia) shown in Figs. 4 and 5 were used as the basis for the interpretation of related pectin spectra presented in Table 3. The intense signal at 172.7 ppm was assigned to the C-6 carbon of COOH group and the intense signal at 101.9 ppm was assigned to the C-1 carbon. The peaks between 60 and 90 ppm are from the carbons of the pyranoid rings (C-2,3,5), and the signal at 80.3 is from the C-4 carbon. The intense signal at 53.7 ppm is assigned to the methyl carbon of the methyl ester groups (COOCH3).

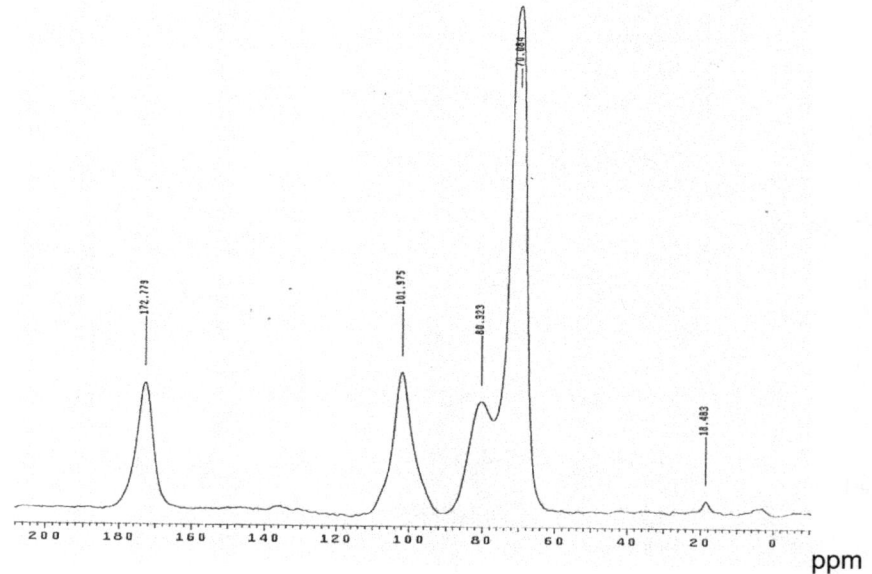

Fig. 4. ^{13}C CP/MAS solid state NMR spectrum of polygalacturonic acid

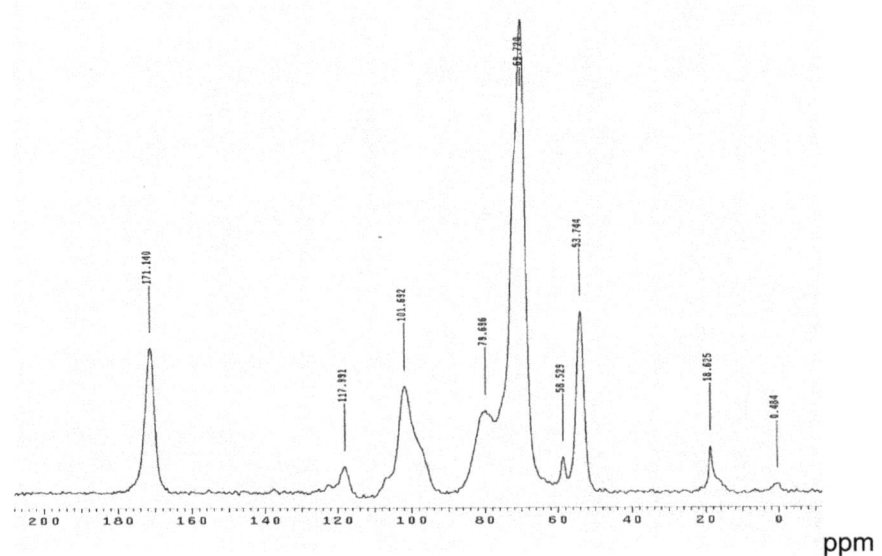

Fig. 5. ^{13}C CP/MAS solid state NMR spectrum of esterified citrus pectin

The spectra of Beauregard sweet potato pectin are shown in Figs. 6-9. It was noticed that the signal of C-6 carbons of galacturonic units of pectin extracted using HCl containing SHMP shifted downfield to 177.25 ppm (Fig. 7), indicating that the galacturonic is in the form of carboxylate anion (COO⁻) [27,36]. The use of NaOH and NaOH containing SHMP for pectin extraction led to the disappearance of the peaks at ~ 53 ppm (Figs. 8 and 9) and this was attributed to saponification of the methyl ester.

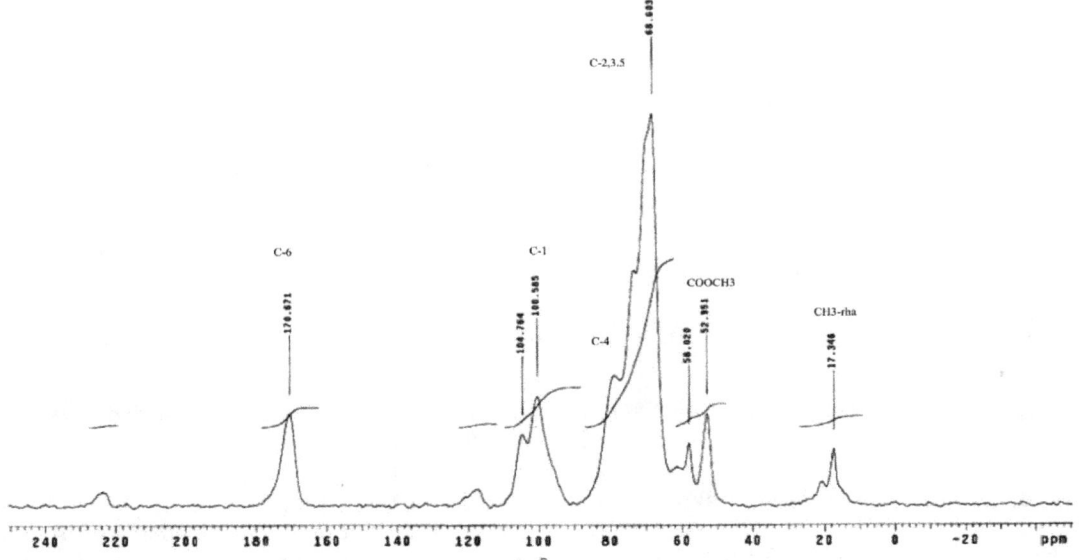

Fig. 6. ^{13} C CP/MAS solid-state NMR spectra of HCl-extracted Beauregard sweet potato pectin

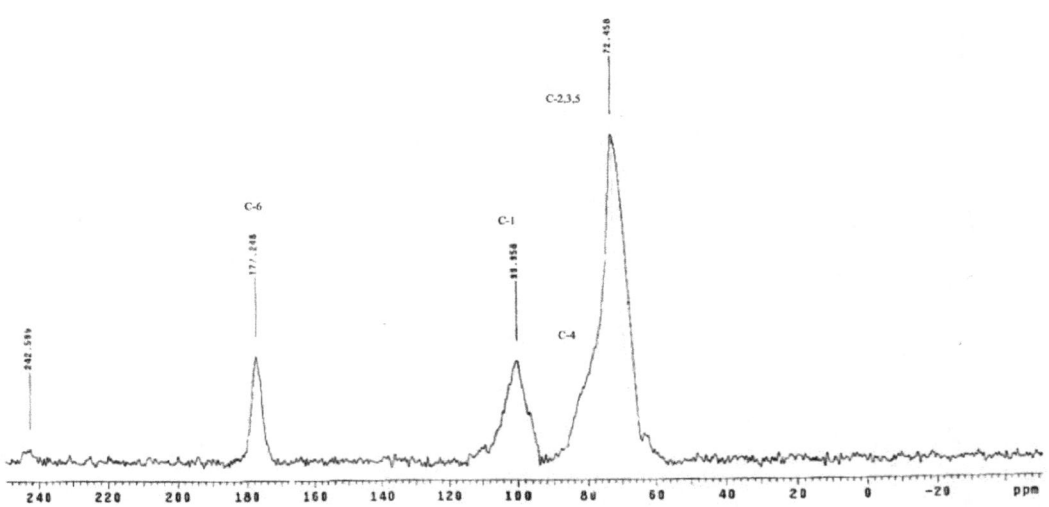

Fig. 7. [13] **C CP/MAS solid-state NMR spectra of HCl containing SHMP-extracted Beauregard sweet potato pectin**

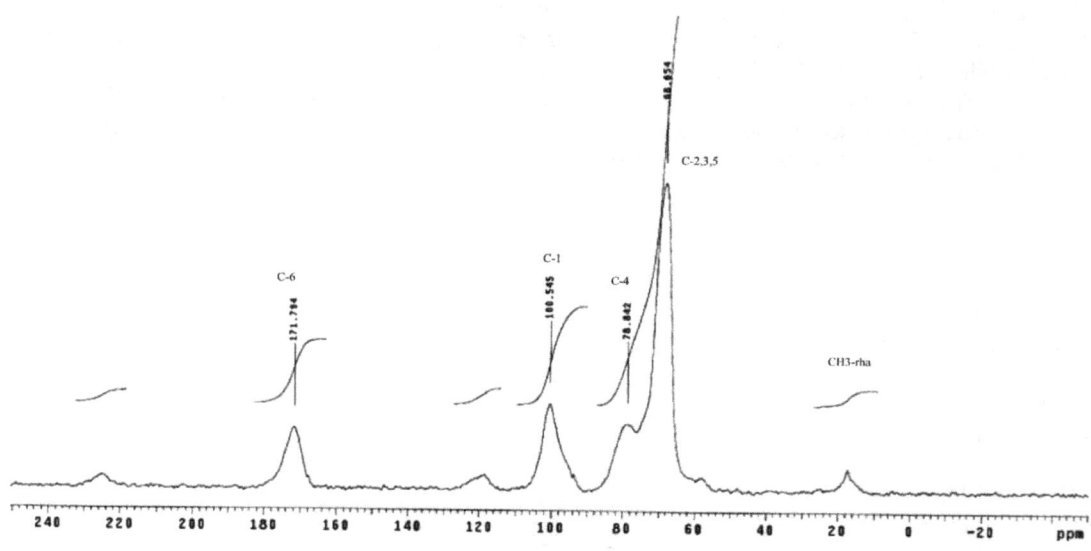

Fig. 8. [13] **C CP/MAS solid-state NMR spectra of NaOH-extracted Beauregard sweet potato pectin**

HCl-extracted pectins have peaks for the C-6 carbon of galacturonic units at ~ 170 – 171 ppm due to relatively high content of the methyl ester groups together with some free carboxyls [27,36,37] In contrast, peaks of the C-6 carbons of galacturonic units of pectin extracted using alkaline or SHMP had peaks between 174-176 ppm. This suggests that the C-6 carbons of these pectins were mostly in the form of carboxylate anion (COO⁻) [27,36].

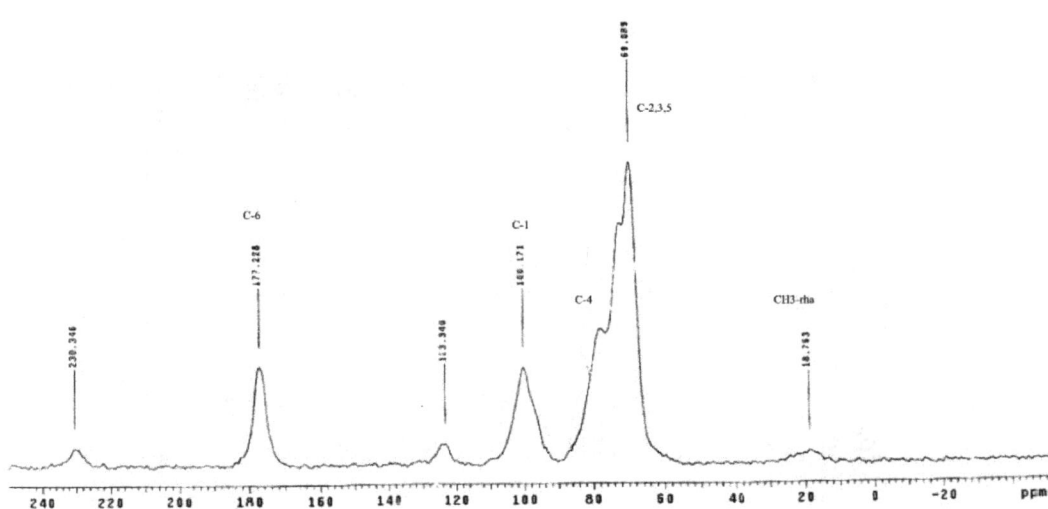

Fig. 9. ¹³C CP/MAS solid-state NMR spectra of NaOH containing SHMP-extracted Beauregard sweet potato pectin

The spectra of pectin extracted using HCl containing SHMP, except those extracted from Beauregard variety, also had resonance signals at ~ 53 ppm representing methyl carbons of the methyl ester (COO$\underline{C}$H₃), whereas those extracted using alkaline or alkaline containing SHMP did not. Interestingly, pectins from Northern Star sweet potato extracted using HCl containing SHMP showed resonance at ~ 20 ppm (Fig. 10). According to Farago and Mahmoud [38] these peaks are from the methyl carbon of acetyl groups (O₂C$\underline{C}$H₃).

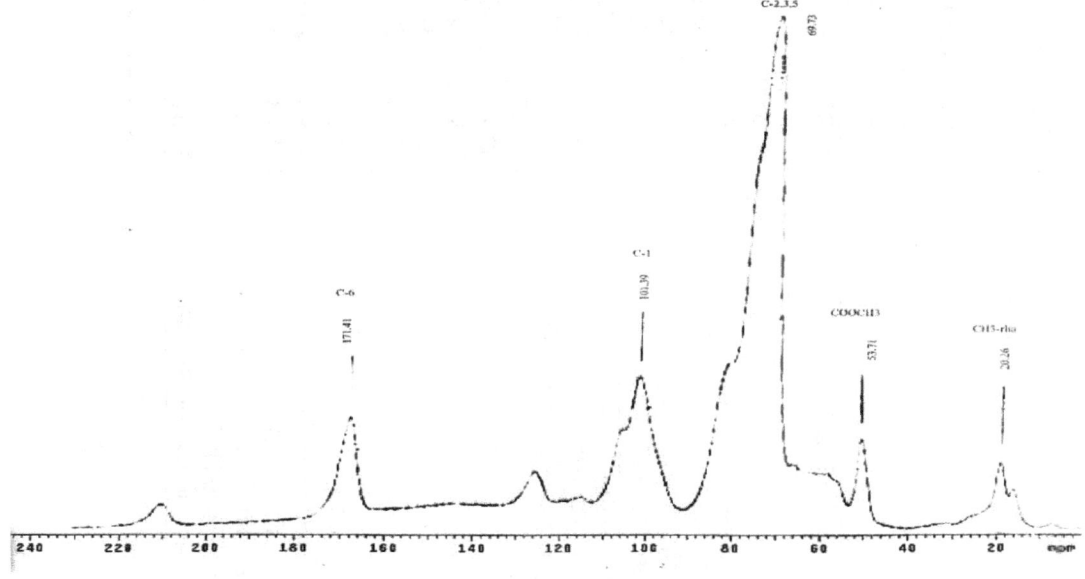

Fig. 10. ¹³C CP/MAS solid-state NMR spectra of HCl containing SHMP-extracted Norhtern Star sweet potato pectin

Table 3. Chemical shift of sweet potato pectin

Sample	C-1	C-2,3,5	C-4	C-6	COOCH₃	O₂CCH₃	CH₃-Rha
			Chemical	shift	(ppm)		
Polygalacturonic acid	101.90	69.81	80.33	172.71	-	-	18.50
Esterified citrus pectin	101.68	69.51	79.00	171.15	53.80	-	-
Beauregard HCl	100.58	68.60	sh	170.67	52.95	-	17.34
Beauregard HCl+SHMP	101.80	71.43	82.49	175.17	-	-	17.42
Beauregard NaOH	100.54	68.65	78.84	171.79	-	-	18.30
Beauregard NaOH+SHMP	99.96	72.46	sh	177.25	-	-	-
Northern Star HCl	102.15	70.06	sh	171.17	53.46	-	-
Northern Star HCl+SHMP	101.39	69.73	sh	171.41	53.71	20.26	-
Northern Star NaOH	102	70.08	80.32	172.77	-	-	18.48
Northern Star NaOH+SHMP	103	73	sh	177	-	-	-
Bis192 HCl	101.37	71.37	sh	171.17	53.59	-	18.48
Bis192 HCL+SHMP	100.26	69.08	sh	176.07	54.81	-	-
Bis192 NaOH	101.57	69.33	sh	172.14	-	-	-
Bis 192 NaOH+SHMP	100.41	69.62	sh	175.88	-	-	18.39
Bis183 HCl	101.43	69.72	sh	171.51	53.59	-	17.99
Bis183 HCl+SHMP	101.53	69.33	sh	171.61	53.93	-	18.28
Bis183 NaOH	100.70	63.81	80.26	175.44	-	-	-
Bis183 NaOH+SHMP	100.99	70.54	sh	175.98	54.17	-	18.19
Starch residue HCl	101.43	69.72	sh	171.51	53.69	-	17.99
Starch residue HCl+SHMP	101.62	69.28	sh	175.41	54.79	-	17.24
Starch residue NaOH	102.21	70.06	sh	173.11	-	-	-
Starch residue NaOH+SHMP	100.51	70.44	sh	176.95	-	-	18.87

sh=shoulder C-2,3,5

Other pectin spectra (Figs. 11-12) show that, in general, HCl-extracted pectin had   peaks for the C-6 carbon of galacturonic units at ~ 170 – 171 ppm due to relatively high content of methyl ester groups together with some free carboxyls [27,36,37]. In contrast, the C-6 carbon signal of pectin extracted using alkaline or SHMP had a resonance signal between 174-176 ppm, which was attributed to sodium galacturonate. It was also observed that the spectra of pectins had intense signals ~ 17 ppm, which corresponds to methyl carbons of rhamnose [27].

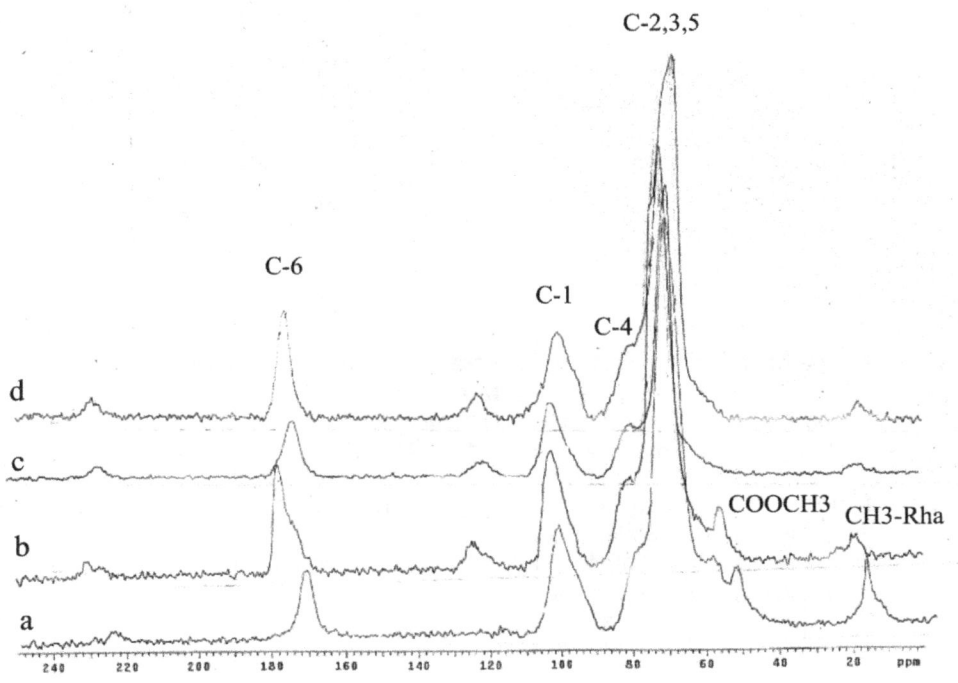

Fig. 11. ^{13}C CP/MAS solid-state NMR spectra of pectin extracted from Bis192 sweet potato

(a) extraction using HCl, (b) extraction using HCl containing SHMP, (c) extraction using NaOH, (d) extraction using NaOH containing SHMP

Fig. 11 shows that the spectra of pectin extracted using HCl as well as HCl containing SHMP had peaks at ~ 53-54 ppm, indicating these pectins contain significant amounts of methyl esters, whereas in pectins extracted using NaOH and NaOH containing SHMP, no COOCH₃ peak was detected due to saponification. Unlike pectin from other sweet potato varieties, it was also observed that carboxyl C-6 carbons of galacturonic units of pectin extracted using NaOH solution were still in the acidic form of COOH, and did not shift downfield to 176 ppm (in the form of COO⁻).

The spectra of pectin from Bis183 sweet potato (Fig. 12) show that extraction using NaOH containing SHMP had a resonance at ~53 ppm, representing the methyl carbon of methyl ester. This phenomenon is different from that extracted using the same method, where this peak was not detected. It was also observed that the peak of the C-4 carbon of pectin extracted using NaOH was more resolved compared to that of the same pectin obtained using different extraction method.

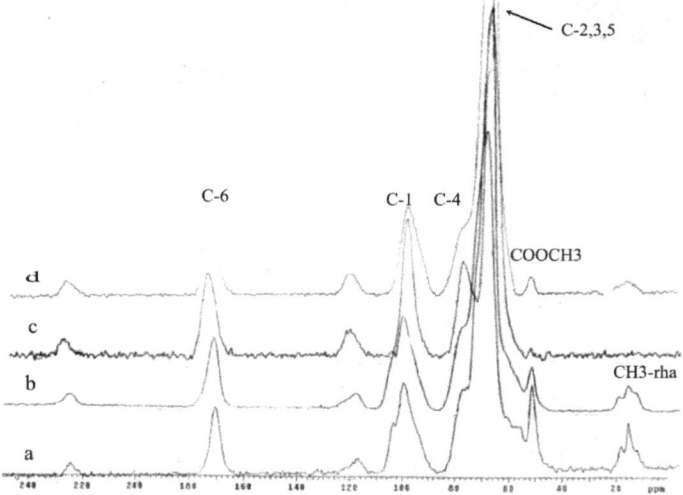

Fig. 12. ¹³ C CP/MAS solid-state NMR spectra of pectin extracted from Bis183 sweet potato.
(a) extraction using HCl, (b) extraction using HCl containing SHMP, (c) extraction using NaOH, (d) extraction using NaOH containing SHMP.

Fig. 13 shows the spectra of commercial starch residue pectin. It was observed that only pectin extracted using HCl had a prominent peak for the methyl carbon of methyl ester and rhamnose whereas others did not, indicating that extraction using SHMP or NaOH has eliminating the methyl ester content of the pectin.

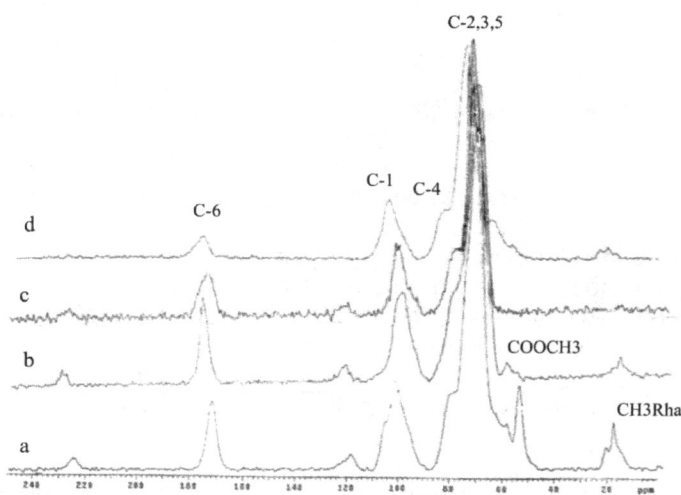

Fig. 13. ¹³C CP/MAS solid-state NMR spectra of pectin extracted from commercial sweet potato starch residue.
(a) extraction using HCl, (b) extraction using HCl containing SHMP, (c) extraction using NaOH, (d) extraction using NaOH containing SHMP

The NMR spectra of the extracted pectins were then used for determining the galacturonic acid content, degree of esterification (DE) and acetylation (DA) [26,27]. Galacturonic acid content was calculated as the ratio of area of the C-6 to the average area of C-1,2,3,4, 5; DE was the ratio of area of COOC$\underline{\text{C}}$H$_3$ to that average area of C-1,2,3,4,5, and DA was the ratio of integral intensity of O$_2$CC$\underline{\text{H}}_3$ carbon to that average area of C-1,2,3,4,5. The area of C-1 was converted to 100 and the other areas were based on that of C-1. The peak area of the pectin samples are presented in Table 4 and their corresponding galacturonic acid and degree of esterification are shown in Table 5.

Table 4. Peak area of the carbons obtained from the NMR spectra

Variety	Extraction method	Area of C-6	Area of C-1	Area of C2,3,4,5	Average area of C-1,2,3,4,5	Area COOCH$_3$
Beauregard	HCl	50.3	100	417.6	103.5	58.8
	HCl+SHMP	46.9	100	492.4	118.5	-
	NaOH	77.6	100	590	138	-
	NaOH+SHMP	64.4	100	437.9	107.5	-
Northern Star	HCl	49.5	100	527.7	125.5	27.5
	HCl+SHMP	56.68	100	587	137	42.5
	NaOH	80.75	100	595.2	139	-
	NaOH+SHMP	93.7	100	506.9	121	-
Bis 192	HCl	34.7	100	536.8	127.3	11.2
	HCl+SHMP	127	100	894	188	18
	NaOH	96.25	100	660.4	152	-
	NaOH+SHMP	75	100	364	92.8	-
Bis 183	HCl	53.5	100	639.7	148	35
	HCl+SHMP	54	100	575	135	41.7
	NaOH	52.9	100	645	149	11
	NaOH+SHMP	57.8	100	628	125.6	-
Starch residue	HCl	47.7	100	639.7	147.9	14
	HCl+SHMP	48.15	100	580.8	136.2	43
	NaOH	35.8	100	279.1	75.8	-
	NaOH+SHMP	86.3	100	639.6	147.9	-

The results (Table 5) show that there were large variations in galacturonic acid content. Galacturonic acid contents were between 27.3 and 80.8%. This is in contrast to Levigne *et al.* [14] where the galacturonic acid contents were reported constant for pectins obtained using a number of extraction conditions. The lowest content of galacturonic acid (27.3%) was found in Bis 192 pectin extracted using HCl and the highest 80.8 was also found in Bis 192 extracted using NaOH containing SHMP. Except for Bis 183, in all varieties, extraction using NaOH plus SHMP gave the highest galacturonic acid contents. It was also noted that the higher the yield of material extracted, the higher the percentage of galacturonic acid. This was attributed to the ability of SHMP to chelate Ca^{2+} and destruction of the alkali-labile linkages such as ester, some glycosidic linkages between methylated galacturonic residues and hydrogen bonds [34], leading to more galacturonic acid being liberated from the cell wall.

Table 5. Galacturonic acid (GA) and degree of esterification (DE), and degree of acetylation (DA)

Varieties	Extraction methods	GA (%)*	DE (%)*	DA
Beauregard	0.1M HCl	48.6	57.0	
	0.1M HCl cont. 0.75%SHMP	39.6	No signal detected	-
	0.05M NaOH	56.0	No signal detected	-
	0.05M NaOH cont.0.75% SHMP	60	No signal detected	-
Northen Star	0.1M HCl	50.9	36	
	0.1M HCl cont. 0.75%SHMP	41.3	31.0	trace
	0.05M NaOH	58.1	No signal detected	-
	0.05M NaOH cont.0.75% SHMP	77.4	No signal detected	-
Bis192	0.1M HCl	27.3	9.0	-
	0.1M HCl cont. 0.75%SHMP	67.6	10.0	-
	0.05M NaOH	63.3	No signal detected	-
	0.05M NaOH cont.0.75% SHMP	80.8	No signal detected	-
Bis 183	0.1M HCl	36.0	23.8	-
	0.1M HCl cont. 0.75%SHMP	40.0	30.9	-
	0.05M NaOH	35.5	7.4	-
	0.05M NaOH cont.0.75% SHMP	46.0	No signal detected	-
Starch residue	0.1M HCl	32.3	9.5	
	0.1M HCl cont. 0.75%SHMP	35.4	32.2	-
	0.05M NaOH	47.2	No signal detected	-
	0.05M NaOH cont.0.75% SHMP	58.2	No signal detected	-

Values are calculated from Table 4. GA = (Area of C-6/ Average area of C1-C5) x 100
DE = (Area of COOCH$_3$/Average area of C1-C5) x 100

Although the galacturonic acid content of the extracted pectin generally was lower than that from commercial pectin as specified by EEC [39], FAO 40] and FCC [41] which is not less than 65%, these results were close to other reports on sweet potato pectin. Sasaki, Kishigami and Fuchigami [42], Noda et al. [20] Salvador et al. [21] reported that the content of galacturonic acid ranged from 0.03 to 0.26% for raw sweet potato, or from 47.1 to 31.3% of sweet potato cell wall materials.

The methyl ester groups were always detected in pectin extracted using HCl, and ranged from 9 to 57%. The highest was in Beauregard and the lowest in Bis 192 sweet potato (Table 5). In contrast, no esterification was detected in the samples extracted with NaOH containing SHMP. Esterification was found in pectins extracted using HCl containing SHMP, except for pectin from Beauregard. HCl containing SHMP-extracted pectins from Northern Star, Bis 192, and Bis 183 had degrees of esterification of 31, 10, 30.9 and 32.2%, respectively. The methyl ester in pectin extracted using NaOH was detected only in Bis 183 sweet potato. These results suggest that alkali extraction had saponified the pectin methyl esters [14, 43] leading to formation of the sodium salt of pectin.

Acetate groups were only detected in pectin from Northern Star and starch residue extracted using HCl containing SHMP, which was too low to quantify. This can be attributed to de-acetylation of pectin by mild acid or alkali hydrolysis during extraction [16].

4. CONCLUSION

The factory starch residue sample and the laboratory-prepared sweet potato starch residue samples both contained considerable amounts of cell wall material (35% to 52%, including pectin between 7 and 30%) and which makes this a feasible raw material for pectin production. The material extracted using HCl, NaOH, and HCl or NaOH containing SHMP, were primarily composed of pectins, since they contained large amounts of galacturonic acid (32% to 80%).

Pectin extracted using HCl as well as HCl containing SHMP contain significant amounts of methyl ester. On the other hand, in pectins extracted using NaOH and NaOH containing SHMP, no $COOCH_3$ peak due to saponification. It was also observed that pectin extracted with HCl, produced galacturonic acid groups (-COOH), whereas extraction with HCl containing SHMP, NaOH, and NaOH containing SHMP, produced galacturonic acid groups in the form of carboxylate anion (COO⁻), except in Bis 192, where the galacturonic units of pectin extracted using NaOH solution were still in the form of COOH.

Extractions with HCl resulted in a fairly high galacturonic acid content, greater degree of esterification, and, in addition, Beauregard appeared to be superior in comparison with other varieties because they had high galacturonic acid content as well as the degree of esterification. In contrast, extraction using NaOH containing SHMP resulted higher pectin yield and galacturonic acid content but very low in esterification. Northern Star had the highest pectin yield (30%), and Bis 192 had the highest galacturonic acid (80%). The overall results show that sweet potato variety had less effect on the yield, galacturonic acid and degree of esterification of pectin compared with the effect of extraction methods.

ACKNOWLEDGEMENT

We would like to thank to Australian Government through AUSAID for funding this research project.

COMPETING INTERESTS

Authors declare that no competing interests exist.

REFERENCES

1. Kertesz ZI. The pectic substances. Interscience Publ. Inc. New York; 1951.
2. Lapasin R, Pricl S. Rheology of Industrial Polysaccharides: Theory and Application. Blackie & Professional, Chapman & Hall, London. p.620. 1995.
3. Rolin C, De Vries J. Pectin. In H.W.S. Chan (Ed.), Food Gels, Chapt.10, pp.401-434. Elsevier Applied Sience, Barking, U.K; 1990.
4. Glicksman M. Gelling hydrocolloids in product applications. In J.V.M. Blanshard and J.R. Mitchell (Ed.). Polysaccharides in Foods. Butterworths, London. 1979;pp.185-204.
5. Pedersen JK. Textural ingredients for food. Business Briefing. Innovative Food Ingredients. 2002;1-4

6.   Yamaguchi F, Shimizu N, Hatanaka C. Preparation and physiological effect of low-molecular weight pectin. Biosci. Biotech. Biochem. 1994;58:679-682.

7.   Tian C, Wang G. Study on the anti-tumor effect of polysaccharides from sweet Potato. Journal of Biotechnology. 2008;136S,S351.

8.   Inngjerdingen M, Inngjerdingen KT, Patel TR, Allen S, Chen XY, Rolstad B. Pectic polysaccharides from Biophytum petersianum Klotzsch, and their activation of macrophages dendritic cells. Glycobiology. 2008;18:1074–1084.

9.   Simpson BK, Egyankor KB, Martin AM. Extraction, purification , and determination of pectin in tropical fruits. J. Food Preserv. 1984;8,63-72.

10.   Matora VA, Korshunova VE, Shkodina OG, Zhemerichkin DA, Ptitchikina NM, Morris ER. The application of bacterial enzymes extraction of pectin from pumpkin and sugar beet. Food Hydrocoll. 1995;9:43-46.

11.   Turquois T, Rinaudo M, Taravel FR, Heyraud A. Extraction of highly gelling pectic substances from sugar beet pulp and potato pulp: Influence of extrinsic parameters on their gelling properties. Food Hydrocoll. 1999;13:255-262.

12.   Gnanasambandam R, Proctor A. Preparation of soy hull pectin. Food Chem. 1999; 65,461-467.

13.   Golovchenko VV, Ovoda RG, Shashkov AS, Odovov YS. Studies of the pectic polysaccharide from duckweed Lemna minor L. Phytochem. 2002;60(1):89-97.

14.   Levigne S, Ralet MC, Thibault JF. Characterisation of pectins extracted from fresh sugar beet under different conditions using an experimental design. Carbohydr. Polym. 2002;49:145-153.

15.   Yapo BM, Robert C, Etienne I, Wathelet B, Paquot M. Effect of Extractions on the yield, purity and surface properties of sugar beet pulp pectin extracts. Food Chem. 2007;100:1356-1364.

16.   Dea ICM, Madden JL. Acetylated pectic polysaccharides of sugar beet. Food Hydrocoll. 1986;1:71-98.

17.   Endress, H.U., and Rentschler, C. Chances and limit for the use of pectin as emulsifier- Part 1. The European Food and Drink Review. Summer. 1999; 49-53.

18.   Byg I, Diaz J, Øgendal LH, Harholt J, Jørgensen B, Rolin C, Svava R, Ulvskov P. Large-scale extraction of rhamnogalacturonan I from industrial potato waste. Food Chem. 2012;131:1207–1216

19.   Yenovsky E. Pectin from sweet potato pulp. Food Industry. 1939;710.

20.   Noda T, Takahata Y, Kumamoto, Nagata T, Shibuya N, Ibaraki. Chemical composition of cell wall material from sweet potato starch residue. Starch/starke. 1994;44:232-236.

21.   Salvador DL, Saganuma T, Kitahara K, Tanoue H, Ichiki M. Monosaccharide composition of sweet potato fiber and cell wall polysaccharides from sweet potato, cassava, and potato analyzed by the high performance chromatography with pulse amperometric detection method. J. Agric. Food Chem. 2000;48:3448-3454.

22.   Salvador LD, Suganuma T, Kitahara K, Fukushige Y, Tanoue H. Degradation of cell wall materials from sweet potato, cassava, and potato by a bacterial protopectinase and terminal sugar analysis of the resulting solubilized products. J. Biosci. Bioeng. 2002;93:64-72.

23. Dongowski G, Stoof G. Investigations on potato pulp as dietary fibre source-composition of potato pulp after influence of pectinases and cellulases and enzymatic degradation of starch. Starch/Starke. 1993;45:230-234.

24. Massiot P, Rouau X, Thibault JF. Characterisation of the extractable pectins and hemicelluloses of the cell-wall of carrot. Carbohydr. Res. 1988;172:229-242.

25. Stolle-Smits T, Beekhuizen JG, Recourt K, Voragen AGJ, Dijk CV. Changes in pectic and hemicellulosic polymers of green beans (Phaseolus vulgaris L.) during industrial processing. J. Agric. Food Chem. 1997;45:4790-4799.

26. Sullivan MJ. Industrial applications of high resolution solid-state 13C NMR techniques. Trends in Anal. Chem. 1987;2:31-37.

27. Sinitsya A, Copikova J, Pavlikova J. 13C CP/MAS NMR spectroscopy in the analysis of pectins. J. Carbohydrate Chem. 1998;17(2):279-292.

28. Steel RGD, Torrie JH. Principles and procedures of statistics.   McGraw-Hill, Singapore. 563 p. 1980.

29. Ryden P, Selvendran RR. Structural features of cell wall polysaccharides of potato (Solanum tuberosum). Carbohydr. Res. 1990;195:257-272.

30. MSDS. Material safety data sheets (MSDS) for DMSO. 2001. Accessed 03 July 2001. Available: http://physchem.ox.ac.uk/MSDS/DM/DMSO.html.

31. Rolin C, De Vries J. Pectin. In H.W.S. Chan (Ed.), Food Gels, Chapt.10, pp.401-434. Elsevier Applied Sience, Barking, U.K; 1990.

32. Rombouts FM, Thibault JF. Feruloyated pectic substances from sugar beet pulp. Carbohydr. Res. 1986;154:177-188.

33. May CD. Industrial pectins: sources, production and applications. Carbohydr. Polym. 1990;12:79-99.

34. Renard CMG., Voragen AGJ, Thibault JF, Pilnik W. Studies on apple protopectin I: extraction of insoluble pectin by chemical means. Carbohydr. Polym. 1990;12:9-25.

35. Doesburg JJ. Pectic substances in Fresh and Preserved Fruit and vegetables. I.B.V.T., Wageningen, The Netherlands; 1965.

36. Catoire L, Goldberg R, Pierron M, Morvan C, du Penhoat CH. An efficient procedure for studying pectin structure which combines limited depolymerization and 13C NMR. Eur. Biophys. 1998; J. 27, 127-136.

37. Jarvis MC, Apperley DC. Chain conformation in concentrated pectic gels: evidence from 13C NMR. Carbohydr. Res. 1995;275:131-145.

38. Farago ME, Mahmoud I. E.A.D.A.W. Metal compounds of pectin. Inorg. Chim. Acta. 1983;80:273-278.

39. E.E.C. Specific criteria of purity for emulsifier, stabilizers, thickeners and gelling agents for use in foodstuffs. European Economic Council. 1978;78/663/EEC.

40. F.A.O. Specifications for identity and purity of thickening agents, anticaking agent, antimicrobial agent, antioxidants and emulsifiers. FAO Food and Nutrition Paper 4. Rome; 1978.

41. FCC. Food Chemical Codex. III Monographs, vol. 215. National Academic Press, Washington D.C; 1981.

42. Sasaki A, Kishigami Y, Fuchigami M. Firming of cooked sweet potatoes as affected by alum treatment. J. Food Sci. 1999;64(1):111-115

43.   Kohn R. Binding of divalent cations to oligomeric fragments of pectin. Carbohydr. Res. 1987;160:343-353.

Effects of Irradiation and Chemical Preservatives on the Microbiological Quality of Refrigerated Fresh-Cut Mangoes

Emmanuel K. Gasu[1*], Victoria Appiah[1], Abraham Adu Gyamfi[2] and Josehpine Nketsia-Tabiri[2]

[1]*School of Nuclear and Allied Sciences Atomic Campus University of Ghana, Ghana.*
[2]*Biotechnology and Nuclear Agriculture Research Institute, Ghana Atomic Energy Commission, Ghana.*

Authors' contributions

This work was carried out in collaboration between all authors. Author EKG designed the study, managed the literature searches, performed the statistical analysis, wrote the protocol, and wrote the first draft of the manuscript. Authors AAG and JNT managed the analyses of the study and Author VA supervised the research. All authors read and approved the final manuscript.

ABSTRACT

Fresh-cut mangoes are nutritious and offer consumers freshness, flavour and convenience. They however have a shorter shelf life compared to whole fruits due to their high susceptibility to microbial contamination. The effects of gamma irradiation and chemical preservatives on the microbiological quality of refrigerated fresh-cut mangoes were evaluated. Well matured fruits of Kent and Keitt varieties sliced into cubes were microbiologically analysed initially to determine counts of total viable cells (TVC), coliforms, *Salmonella sp.*, *Staphylococcus aureus* and *Escherichia coli*. The samples were subjected to various irradiation doses (0, 1.0, 1.5, 2.0 and 2.5 kGy) and chemical preservatives (sucrose, citric acid, sodium benzoate and a combination of these chemicals in equal proportions) and stored at 6°C and 10°C for 15 days. TVC was subsequently estimated at 3-day intervals for the treated samples. TVC was estimated as 3.53 ± 0.25 and 4.86 ± 0.38 $\log_{10}$cfu/g for the Kent and Keitt varieties respectively. No coliforms *Salmonella* sp., *E. coli* or *S. aureus* were detected in both varieties. Irradiation at doses of 1.5 kGy to 2.5 kGy in combination with storage at 6°C was able to eliminate

Corresponding author: Email: ekgasu@yahoo.co.uk

all viable cells after 9 days compared to 12 days of storage at 6°C in the case of chemical preservatives. Irradiation is more effective and ideal compared to chemical preservatives in improving the microbiological quality and therefore extending the shelf life of refrigerated fresh-cut mangoes.

Keywords: Irradiation; fresh-cut mangoes; total viable cells; microbial load, shelf life.

1. INTRODUCTION

Generally, fruits form an important natural staple food and are excellent sources of minerals, vitamins, fibres, anti-oxidants and enzymes. Fresh-cut mangoes are minimally processed fruits and therefore offer consumers freshness, flavour and convenience apart from their nutritional value Lamikanra [1]. The rising public awareness of their health benefits have led to increased production and consumption in recent years Ragaert et al. [2]. Accordingly, sale of fresh-cut fruits such as pawpaw, watermelon and pineapple wrapped in polyethylene has significantly increased in Ghana due to their convenience. Despite the health benefits, fresh-cut fruits have a shorter shelf life compared to whole fruits due to a number of factors. They are highly susceptible to microbial contamination and spoilage, require specific storage conditions and undergo gradual changes in quality Chien et al. [3]. Contamination of fresh-cut fruits with disease-causing microorganisms including human pathogens is common EC [4]. Consequently, there have been increased outbreaks of food poisoning and food infection diseases associated with their consumption, thus posing food safety risks Allong et al. [5], Herndon [6] and UFPA [7].

Several procedures are used to ensure microbiological stability and acceptable hygienic quality of fresh-cut fruits. Production practices such as Good Agriculture Practices (GAP) and Good Hygienic Practices (GHP) in combination with Hazard Analysis and Critical Control Point (HACCP) have been utilized to minimize the level of microbial contamination of fresh-cut fruits. Additionally, appropriate temperature and moisture management, vacuum and modified atmosphere packaging, irradiation and the use of chemical preservatives have been employed to control proliferation of microorganisms in fresh-cut fruits. Irradiation, which serves as a Critical Control Point in HACCP application, has been utilized in controlling microbial contamination of foods over several decades. Studies have consistently shown that irradiation effectively kills bacterial pathogens on fresh and fresh-cut produce Smith and Pillai [8], Niemira and Fan [9]. For several years, a number of chemical compounds such as the chlorine-based preservatives have also been used to reduce bacterial populations on fruits especially before processing or during pre- and post-cutting operations Gómez-López et al. [10] and Gil et al. [11].

In spite of the widespread utilization of irradiation and chemical preservatives for shelf life extension of fresh-cut produce at the international level, there is scanty information on the potential application of these technologies in Ghana. The objective of this study was therefore to investigate the effect/s of gamma irradiation and chemical preservatives on the microbiological quality of fresh-cut mangoes during refrigerated storage.

2. METHODS

2.1 Sample Collection and Preparation

Fifty each of matured Kent and Keitt varieties of mangoes were purchased freshly harvested from farms in Somanya in the Eastern Region of Ghana and transported in plastic crates to the laboratory. The samples were washed in 10% sodium hypo-chlorite solution and rinsed in sterile distilled water. Samples weredivided into three batches for preliminary microbiological quality assessment, irradiation and treatment with chemical preservatives. They were then peeled, seeds removed and processed into cubes.

2.2 Irradiation of Samples

Samples of the fresh-cut mangoes were packaged into polyethylene terephthalate (PET) jars for irradiation at various doses (0, 1.0, 1.5, 2.0, and 2.5 kGy). Irradiation of the samples was carried out in air at a dose rate of 2.42 kGy/h using a Cobalt-60 source at the Gamma Irradiation Facility of the Ghana Atomic Energy Commission. The absorbed dose was determined by using Lithium fluoride photo-fluorescent film dosimeter (SUNNA Dosimeter System, UK).

2.3 Chemical Preservation of Fresh-Cut Mangoes

Samples of fresh-cut Kent and Keitt mangoes were treated with the following preservatives: 30 g/L Sucrose, 3.0 g/L of Citric acid, 2.0 g/L Sodium benzoate and a combination of these three preservatives in equal proportions as the individual chemicals.

2.4 Storage of Chemically Preserved Fresh-Cut Mangoes

The irradiated and chemically-preserved samples and their controls were stored in a refrigerator (IGNIS Model: RWN130) set at 6°C and 10°C for 15 days each. A portable laboratory thermometer was kept in the refrigerators to monitor the temperatures throughout the storage period.

2.5 Microbiological Analysis

All the three batches of fresh-cut mangoes were microbiologically analyzed to determine the population of indicator and pathogenic microorganisms. In the case of samples for irradiation and chemical preservation, microbiological analysis was undertaken before and after treatment. Total viable cells, total coliform counts, counts of *Salmonella sp.*, *Staphylococcus aureus* and *Escherichia coli* were determined for all the three batches of samples. For each sample, 10 g was weighed into 90 ml Peptone water diluent (0.1% peptone and 0.5 NaCl) and stirred on a mechanical shaker (Junior Obit Shaker, Lab-Line Instrument, USA) for 30 minutes and serially diluted up to 106. One milliliter aliquots from each dilution were dispensed into petri dishes and about 15 ml of the appropriate media was added. Total viable cells were determined on Plate Count Agar (Oxoid, England). Total Coliform counts were determined on Violet Red Bile Agar (Oxiod, England) and *Staphylococcus aureus* was estimated on Baird- Parker (BP) agar (Oxoid, England). *E. coli* was determined on Eosin methylene blue (EMB). All determinations were undertaken in duplicate and three separate experiments were conducted.

All samples were incubated at 37°C for 24 hours and observed for colonies. Plates that had between 30-300 colonies were selected for the determination of colony forming units per gram (cfu/g) using a colony counter (Staurt Scientific, UK). The number of cfu/g was calculated by multiplying the number of bacteria by the dilution factor.

2.6 Data Analysis

The mean count of total viable cells were calculated and transformed into logarithms. The mean log10(x) values and standard deviations (SD) were calculated on the assumption of a log normal distribution.

3. RESULTS

3.1 Effect of Irradiation on Total Viable Cells of Refrigerated Fresh-Cut Mangoes during Storage

The microbiological quality of fresh-cut mangoes indicated that Kent and Keitt varieties have total viable cells (TVC) of 3.53 ± 0.25 and 4.86 ± 0.38 log10 cfu/g respectively. No coliforms, *E. coli or Staphylococcus aureus* were detected in both varieties. The number of TVC of Kent and Keitt varieties initially increased on the day three for the non-irradiated control sample but decreased over the storage period (Tables 1 and 2). The number of TVC for fresh-cut Kent and Keitt varieties decreased with increasing storage time and increasing irradiation dose at both 6°C and 10°C (Tables 1 and 2). No viable cells were detected in the irradiated samples at both temperatures after storage for 12 days and beyond. With regards to samples irradiated at 1.5 kGy to 2.0 kGy, no viable cells were detected after storage for 9 days and beyond at 6°C and for 12days and beyond at 10°C. Samples irradiated at 2.5 kGy and stored at 6°C had no viable cells beyond the first day of storage for the Kent variety and after storage for 9 days and beyond for the Keitt variety.

In the case of samples irradiated at 2.5 kGy and stored at 10°C, no viable cells were detected after storage for 6 days and beyond for the Kent variety; and after storage for 3 days and beyond for the Keitt variety. The results seem to indicate that whiles there were no appreciable differences in the effect of irradiation on the number of TVC of the two varieties at the storage temperature of 10°C, irradiation had a greater impact on the TVC of the fresh-cut Kent variety compared to the Keitt variety at a storage temperature of 6°C (Tables 1 and 2).

3.2 Effect of Chemical Preservatives on Total Viable Cells of Refrigerated Fresh-Cut Mangoes During Storage

There were generally non-uniform but gradual decreases in the number of TVC of the samples of fresh-cut Kent and Keitt varieties treated with chemical preservatives over the storage period of 15 days at 6°C and 10°C (Tables 3 and 4). Compared to the untreated control, Citric acid (3 g / L) and Sodium benzoate (2 g/L) reduced the number of TVC of the Kent variety by more than 2 log cycles after 15 days of storage at both temperatures. The mixture of chemical preservatives (Sucrose/Citric acid /Sodium benzoate) eliminated all TVC of both Kent and Keitt samples after 9 days and beyond after storage at 6°C and 10°C. With the exception of Sucrose, all the chemical preservatives were more effective in reducing the number of TVC of the Kent variety compared to the Keitt variety.

Table 1. Effect of irradiation on total viable cells of refrigerated fresh-cut Kent mango

Storage temperature	Storage day	Irradiation dose (kGy)				
		0	1.0	1.5	2.0	2.5
6°C	0	3.55 ± 0.35	3.43 ± 0.49	3.10 ± 0.99	3.13 ± 0.07	1.20 ± 0.00
	3	4.80 ± 0.00	2.97 ± 3.53	2.55 ± 0.00	2.37 ± 0.00	-
	6	3.60 ± 0.31	2.97 ± 3.53	2.00 ± 0.56	1.50 ± 0.00	-
	9	3.60 ± 0.03	2.97 ± 3.53	-	-	-
	12	2.95 ± 0.78	-	-	-	-
	15	-	-	-	-	-
10°C	0	3.53 ± 0.25	3.45 ± 0.07	3.43 ± 1.79	3.37 ± 0.58	2.45 ± 0.07
	3	4.93 ± 0.58	3.47 ± 0.84	2.17 ± 0.08	2.25 ± 0.64	1.40 ± 0.00
	6	3.53 ± 0.30	3.20 ± 0.27	2.10 ± 0.69	1.80 ± 0.14	1.10 ± 0.00
	9	3.45 ± 0.12	2.70 ± 0.00	2.00 ± 0.00	1.00 ± 0.00	-
	12	3.45 ± 0.77	-	-	-	-
	15	3.45 ± 0.00	-	-	-	-

Values are means ±S.D counts of log_{10} cfu/10g; (-) = no colonies observed at minimum detection level of 1.0 log_{10}

Table 2. Effect of irradiation on total viable cells of refrigerated fresh-cut Keitt mango

Storage temperature	Storage day	Irradiation dose (kGy)				
		0	1.0	1.5	2.0	2.5
6°C	0	4.70 ± 1.54	4.30 ± 0.38	4.30 ± 1.54	3.50 ± 1.41	3.25 ± 0.05
	3	5.10 ± 0.30	3.60 ± 0.28	3.30 ± 1.08	3.05 ± 1.76	2.90 ± 0.05
	6	3.30 ± 0.55	2.97 ± 1.16	2.80 ± 0.98	2.50 ± 0.70	2.15 ± 0.00
	9	2.30 ± 0.00	2.55 ± 0.00	-	-	-
	12	2.28 ± 0.23	-	-	-	-
	15	2.28 ± 0.12	-	-	-	-
10°C	0	4.83 ± 0.24	4.83 ± 0.81	4.50 ± 0.08	3.17 ± 0.15	2.60 ± 0.79
	3	3.90 ± 0.44	3.83 ± 0.58	2.87 ± 0.71	1.93 ± 0.37	1.33 ± 0.58
	6	3.90 ± 0.82	2.95 ± 0.07	2.37 ± 0.46	1.45 ± 0.64	-
	9	3.33 ± 0.32	2.20 ± 0.00	2.00 ± 0.00	1.30 ± 0.00	-
	12	3.20 ± 0.71	-	-	-	-
	15	3.20 ± 0.11	-	-	-	-

Values are means ±S.D counts of log_{10} cfu/10g; (-) = no colonies observed at minimum detection level of 1.0 log_{10}

Table 3. Effect of chemical preservatives on total viable cells of refrigerated fresh-cut Kent mango

Storage temperature	Storage day	Chemical preservatives (g/L)				
		Control	Sucrose (S)	Citric acid (CA)	Na benzoate (SB)	(S/CA/SB)
	0	3.60 ± 0.00	3.86 ± 1.59	3.63 ± 0.00	3.45 ± 0.48	4.50 ± 0.15
	3	3.98 ± 0.28	3.65 ± 0.92	2.83 ± 0.00	3.30 ± 1.69	1.30 ± 0.00
	6	3.85 ± 0.28	3.20 ± 0.28	1.62 ± 0.73	2.26 ± 1.69	1.30 ± 0.00
6°C	9	3.83 ± 0.28	2.20 ± 0.28	1.60 ± 0.70	1.15 ± 0.00	-
	12	3.52 ± 0.28	-	-	-	-
	15	3.20 ± 0.28	-	-	-	-
	0	4.75 ± 0.35	4.50 ± 0.35	4.40 ± 1.55	4.60 ± 0.10	3.50 ± 0.00
	3	3.97 ± 0.51	3.75 ± 1.06	3.60 ± 1.38	3.57 ± 1.61	1.00 ± 0.00
	6	3.97 ± 1.91	3.23 ± 1.01	3.40 ± 0.98	2.95 ± 0.07	1.80 ± 0.00
10°C	9	3.77 ± 1.70	3.23 ± 1.59	3.00 ± 0.00	1.30 ± 0.00	-
	12	3.73 ± 0.04	2.90 ± 1.32	1.90 ± 0.00	1.30 ± 0.00	-
	15	2.65 ± 0.49	2.50 ± 1.20	-	-	-

Values are means ±S.D counts of log_{10} cfu/10g; (-) = no colonies observed at minimum detection level of 1.0 log_{10}; S = 30g/L; CA = 3g/L, SB = 2g/L

Table 4. Effect of chemical preservatives on total viable cells of refrigerated fresh-cut Keitt mango

Storage temperature	Storage day	Chemical preservatives (g/L)				
		Control	Sucrose (S)	Citric acid (CA)	Na benzoate (SB)	(S/CA/SB)
	0	4.90 ± 0.00	4.35 ± 1.62	4.15 ± 0.21	4.69 ± 0.45	4.00 ± 0.00
	3	4.85 ± 0.00	2.82 ± 0.97	3.72 ± 0.30	4.37 ± 1.98	4.02 ± 0.25
	6	3.90 ± 0.00	2.75 ± 2.47	3.00 ± 0.00	3.10 ± 0.00	1.80 ± 0.42
6°C	9	3.20 ± 0.00	2.50 ± 0.00	2.32 ± 0.00	2.00 ± 0.00	-
	12	3.20 ± 0.31	-	-	-	-
	15	3.19 ± 0.05	-	-	-	-
	0	4.53 ± 0.28	4.73 ± 1.35	4.53 ± 1.05	3.59 ± 0.40	3.45 ± 0.35
	3	3.75 ± 0.28	4.13 ± 2.47	4.10 ± 0.65	3.55 ± 0.07	2.90 ± 0.00
	6	3.96 ± 1.29	2.97 ± 1.29	3.40 ± 0.36	3.37 ± 0.23	2.48 ± 0.17
10°C	9	3.90 ± 0.69	2.73 ± 1.50	3.05 ± 0.76	2.70 ± 0.87	-
	12	3.00 ± 0.00	2.50 ± 0.00	3.00 ± 0.00	2.50 ± 0.00	-
	15	2.50 ± 0.00	2.10 ± 0.00	2.90 ± 0.00	-	-

Values are means ±S.D counts of log_{10} cfu/10g;(-) = no colonies observed at minimum detection level of 1.0 log_{10};S = 30g/L; CA = 3g/L; SB = 2g/L

4. DISCUSSION

4.1 Effect of Irradiation on Total Viable Cells of Refrigerated Fresh-Cut Mangoes during Storage

The microbiological quality of fresh-cut mangoes was acceptable largely because they did not contain coliforms and other potential pathogens. Since total viable cells (TVC) represent mainly spoilage microorganisms, their population is significant and therefore largely determines the microbiological quality of fresh-cut mangoes. This finding may be the result of good manufacturing procedures employed in processing the samples as has also been observed elsewhere EC [4].

Irradiation had an impact on microbiological quality of fresh-cut mangoes since the number of TVC reduced over the storage period for both Kent and Keitt at the storage temperatures of 10°C and 6°C. The study has shown that the microbiological quality of cut mangoes can be improved by treatment with ionizing radiation at doses less than 2.5kGy. This agrees with Patterson and Stewart [12] and Farkas et al. [13] that irradiation of pre-packaged vegetables and ready to eat meals including fruits at 2.0 kGy and 3.0 kGy reduced the total viable cells to levels below the detection limit of 100 cells/g and counts did not increase significantly during storage at 5°C. In a related study on apple and pear jam, Mossel [14] and Abadias et al. [15] reported an increase in microbial load within the first 3 days of storage at 10°C and there after a reduction.Chantanawarangoon [16] reported rapid increases in both the total microbial and yeast and mould counts in the control experiment of mango cubes after 4 days of storage at 5°C. When storage was extended up to 10 days, there were no significant differences in total microbial and yeast and mold counts of mango cubes stored at 5°C except the control, which had higher microbial population as was also found in this study.

Temperature seems to contribute to the low microbial counts as seen in Kent and Keitt samples irradiated and stored at 6°C compared to storage at 10°C. Several studies have confirmed that good temperature management lowers the rate of physiological activities of fruits and therefore prolongs the shelf life of fruits (Allong et al. [5]; Farkas et al. [13]; Farzana [17]. In a study on non-irradiated fresh-cut 'Carabao' mangoes, Izumi et al. [18] noted that storage at 5°C maintained the quality of mango cubes and extended the shelf-life between 4 to 6 days. Furthermore, Dea et al. [19], experimenting with fresh-cut Kent mangoes found that storage at 12°C could only extend the shelf-life between 3 to 4 days whiles at 5°C the shelf life was extended to between 5 to 6 days In this work irradiation has been shown to reduce the number of TVC and therefore the microbial load, thus extending the shelf life of fresh-cut mangoes. The action of radiation is through the radiolysis of water which generates free radicals, which in turn attack the DNA and other organelles in microorganisms, thereby causing their inactivation and eventual elimination Fan et al. [20].

4.2 Effect of Chemical Preservatives on Total Viable Cells of Refrigerated Fresh-Cut Mangoes during Storage

Chemical preservatives are only protective because they are limited to the surface of the fresh cut products. They do not penetrate the tissues of the products and are therefore unable to inactivate the DNA and therefore unable to completely eliminate the total viable cells. Chemical preservatives alter the pH of the medium as well as that of the bacterial membrane resulting in exposure of viable cells to high osmotic concentration Gabrielson [21]. In this study, the chemical preservatives were effective in eliminating the viable cells of the fresh-cut mangoes especially at the storage temperature of 6^0C after 12 days of storage and beyond. It is important to also note that in the combination preservatives, the chemicals acted synergistically in the mixture to effectively eliminate the viable cells of the fresh-cut mangoes after storage for 9 days and beyond and therefore improved the microbiological quality of the fresh-cut mangoes.

Microorganisms play a very important role in determining the microbiological quality and therefore shelf-life and safety of food products. The impact of chemical preservatives on the microbiological quality of fresh-cut mangoes had been demonstrated by the results of this study. Several chemical compounds have been used to reduce bacterial populations on fruits and they are still the most widely used treatments, either before processing or during pre-cutting and post-cutting operations. Treating fresh-cut mangoes with chemical preservatives can improve shelf-life and microbiological safety by destruction of spoilage microorganisms. In a related study, the shelf life, quality and microbiological safety of minimally processed mangoes were improved by the use of chemical preservatives Ragaert et al. [2] In another study on a closely related fruit, jackfruit (Artocarpusheterophyllus L) Saxena et al. [22] also reported shelf life extension in fresh-cut jackfruit pretreated with Sodium benzoate (0.045% w/v) in combination with proper surface sanitisation.

The results of the present study seem to suggest that temperature contributed to the low microbial counts as seen in Kent and Keitt samples stored at 6°C as compared to storage at 10°C. Farzana [17] noted that by reducing storage temperature as was also the case in this study, the rates of metabolism in microbes and fruit tissues can be slowed, thus extending shelf life. It has also been established that good temperature management is very important in prolonging the shelf life of mangoes, and that 2 to 5°C is the optimum storage temperature range since storage at 0°C for more than 10 days caused chilling injury Kader [23].

4.3 Comparative Effects of Irradiation and Chemical Preservatives on Microbiological Quality of Refrigerated Fresh-Cut Mangoes during Storage

On the basis of comparing the impact of irradiation and chemical preservatives on microbiological quality of fresh-cut mangoes, the study revealed the effectiveness of irradiation in combination with refrigerated storage in reducing the population of spoilage microorganisms i.e. total viable cells. This study has shown that irradiation doses of 1.5kGy to 2.5kGy in combination with storage at 6°C were able to eliminate all viable cells after 9 days compared to 12 days of storage in the case of chemical preservatives. The role of irradiation in eliminating microorganisms from food is well documented Smith and Pillai [8], Niemira and Fan [9]. The radiolysis of water by ionizing radiation generates free radicals, which in turn, destroy the DNA and other organelles in microorganisms and thereby inactivating and eliminating them J Am Diet Assoc. [24] and Fan et al. [20]. The capacity of irradiation to eliminate microorganisms in foods without accumulation of residues makes the technology suitable as a food processing tool. Chemical preservatives have been identified as incapable of completely eliminating microorganisms on fresh produce, but rather act as osmotic agents which change the membrane potentials of the microorganisms through enzymatic reactions thereby inactivating them Abadias et al. [15]. Aside not being as effective as irradiation in eliminating viable cells from the fresh-cut mangoes, chemical preservatives have also been known to leave residues in foods, thus creating food safety concerns for consumers.

This study has revealed irradiation as more effective and ideal compared to the use of chemical preservatives in improving the microbiological quality and therefore extending the shelf life of refrigerated fresh-cut mango.

Subsequent studies will explore the impact of irradiation doses of 1.5 to 2.5 kGy on the sensory qualities of refrigerated fresh-cut mangoes.

ACKNOWLEDGEMENTS

I wish to acknowledge Dr F.K Saalia, Department of Nutrition and Food Science, University of Ghana. Ms Theodosia Adom, Radiological and Medical Sciences Research Institute Ghana Atomic Energy Commission for their encouragement during this research

COMPETING INTERESTS

Authors have declared that no competing interests exist.

REFERENCES

1. Lamikanra O. Preface. In Lamikanra. O eds, Fresh-cut fruits and vegetables. Science technology and market Boca Raton.FL:CRC Press; 2002.
2. Ragaert P, Verbeke W, Devlieghere F, De bevere J. Consumer perception and choice of minimally processed vegetables and packaged fruits. Food Quality and Preference. 2004;15:259-270.
3. Chien PJ, Sheu F, Yang FH. Effects of edible chitosan coating on quality and shelf life of sliced mango fruit. Journal of Food Engineering. 2007;78:225-229.
4. European Commission. Risk Profile on the microbiological contamination of fruits and vegetables, Report of the Scientific Committee on Food; 2002.
5. Allong R, Wickham LD, Mohammed M. The effect of cultivar, fruit ripeness, storage temperature and duration on quality of fresh-cut mango. Acta Hort. 2000;509:487-494.
6. Herndon M. FDA Issues Final Guidance for Safe Production of Fresh-Cut Fruits and Vegetables FDA Press release March 12th 2007. http://naturalstandard.com/news/news200703022.asp. Retrieved: Oct.2010.
7. United Fresh Produce Association (UFPA). U.S. Department of Health and Human Services Food and Drug Administration Center for Food Safety and Applied Nutrition February 2008.
8. Smith JS, Pillai S. Irradiation and food safety, An IFT Scientific Status Summary: Food Technology. 2004;58(11):48-55.
9. Niemir BA, Fan X. Low-dose irradiation of fresh and fresh-cut produce: Safety, sensory and shelf life. In" Food Irradiation Research and Technology," ed. Sommers, C. and Fan X, 2005;169-181. Blackwell Publishing and Institute of Food Technologists, Ames, Iowa.
10. Gómez-López VM, Rajkovic A, Ragaert P, Smigic N, Devlieghere F. Chlorine dioxide for minimally processed produce preservation: a review. Trends in Food Science and Technology. 2009;20:17-26.
11. Gil MI, Kader AA. Fresh-cut fruits and vegetables, p. 475-504, in: Tomas-Barberan, F. and MI Gil (editors); 2009. Improving the health-promoting properties of fruits and vegetable products CRC Press, Boca Raton, FL.
12. Patterson MF, Stewart EM. Effect of Gamma Irradiation on the Shelf-life and Nutritional Quality of ready-made meals IAEA-DOC, 2003.
13. Farkas J, Polyák-Fehér K, Mohácsi Farkas C, Mészáros L, Andrássy E, Sáray T. Studies on irradiation of pre-packaged prepared vegetables and improvement of microbiological safety of some sous-vide meals by gamma radiation. IAEA. 2003;27.

14. Mossel DAA. Irradiation: An effective mode of processing food for safety, IAEA –SM-271/80; 1980.

15. Abadias M, Usall J, Oliveira M, Alegre I, Vinas I. Efficiency of neutral electrolyzed water (NEW) for reducing microbial contamination on minimally-processed vegetables. International Journal of Food Microbiology. 2008;123:151-158.

16. Chantanawarangoon S. Quality maintenance of fresh-cut mango cubes, M.S. Thesis in Food Science, University of California at Davis; 2000.

17. Farzana Panhwar. Postharvest technology of mango fruit, development, physiology, pathology and marketing in Pakistan; 2005.

18. Izumi H, Nagatomo T, Tanaka C, Kanlayanarat S. Physiology and quality of fresh-cut mango is affected by low O2 controlled atmosphere storage, maturity and storage temperature. Acta Hort. 2003;600:833-838.

19. Dea S, Brecht JK, Nunes MCN, Baldwin EA. Incidence of chilling injury in fresh-cut 'Kent' mangoes. Hort Science. 2008;43:1148.

20. Fan X, Niemira A, Brendan, Prakash Anuradh. Irradiation of fresh fruits and vegetables. Journal of Food technology. 03:08. 2008;62(3):36.

21. Gabrielson Jenney. Assessing the toxic impact of chemicals using bacteria; 2004. http://hdl.handle.net/10616/39303, Retrieved: Oct. 2010.

22. Saxena A, Bawa AS, Raju PS. Use of modified atmosphere packaging to extend shelf-life of minimally processed jackfruit (*Artocarpus heterophyllus* L.) bulbs. Journal of Food Engineering. 2008;87:455–466.

23. Kader AA. Fresh-cut mangoes as a value-added product (literature review and interview) Postharvest Horticulture Consultant, Kader Consulting Service, USA; 2008.

24. Journal of American Dietetic Assoc. Food irradiation, Position of ADA, 2000;100:246-253.

11

Physicochemical Characteristics of "Gari" Semolina Enriched with Different Types of Soy-melon Supplements

M. O. Oluwamukomi[1*] and I. A. Adeyemi[2]

[1]Department of Food Science and Technology, Federal University of Technology, Akure, Nigeria.
[2]Bells University of Technology, Ota, Nigeria.

Authors' contributions

This work was carried out in collaboration between all authors. Authors MOO and IAA designed the study, author MOO performed the statistical analysis, wrote the protocol, and wrote the first draft of the manuscript. Author MOO managed the analyses of the study. Author MOO managed the literature searches. All authors read and approved the final manuscript.

ABSTRACT

The effects of enrichment of "gari" semolina with three different types of soy-melon protein supplements during toasting of "gari" were studied. Three protein supplements (Full fat, Defatted and Milk residue) were toasted together separately with the grated, dewatered and sifted cassava mash after fermentation (soak-mix method). After toasting and cooling, the samples were subjected to physico-chemical analyses. Results showed that enrichment increased the protein, fat and ash contents, and the pH values, while the hydrocyanic acid content, titratable acidity reduced generally. Enrichment increased the protein content from 2.81% in the control gari to a range of 15.3% - 23.5% in the enriched samples. The fat increased from 3.24% to a range of 4.13% - 13.50%; while the ash content increased from 1.18% to a range of 1.96% to 3.47%. Hydrocyanic acid was significantly ($P \leq 0.05$) reduced from 13.5mg/kg to a range of 6.70mg/kg -12.5mg/kg in the enriched products. The pH increased from 3.62 to a range of 4.86 - 5.25 while the acidity correspondingly reduced from 0,46 in the control gari to a minimum value of 0.36% lactic

Corresponding author: Email: mukomi2003@yahoo.com;

acid in the sample toasted together with defatted soy-melon meal. From the result it could be concluded that enrichment improved the nutrient quality of "gari" especially the protein, fat and ash contents. It also reduced the hydrocyanic acid content, thereby producing "gari" of higher quality and better safety. The acidity of the enriched samples was however reduced thus lowering the sourness of "gari". This may be an advantage for people who are not used to the sour taste of "gari". Of all the three soy-melon "gari" samples, the sample enriched with defatted supplement had been shown to have the highest protein and ash contents, the lowest crude fat and acidity than other enriched gari samples. It had also been shown to have better wettability, water holding capacity, and ability to disperse in water. It also had better swelling and reconstitution indices than "gari" enriched with full fat and milk residue supplements.

Keywords: Enrichment; soy-melon gari; hydrocyanic acid.

1. INTRODUCTION

"Gari" is a fermented, dewatered and toasted starchy granule from cassava which is widely consumed all over West Africa and in Brazil where it is known as 'farinha de manioca' [1]. "Gari" is one of the most popular forms in which cassava (Manihot esculenta Crantz) also known as manioc is consumed in Nigeria and some other parts of West Africa [2]. It is a major component of everyday diet in Nigeria providing about 11.835kJ/person/day [3]. Micronutrient deficiencies in the diet of African countries have been implicated as a national problem which could lead to nutritional insecurity if not adequately tackled. Cassava is therefore recognized as a potential vehicle for micronutrient intervention in Africa [4]. However, Cassava from which this important item of food is produced is low in protein and deficient in essential amino acids. The crude protein content of locally produced "gari" is 1.03% and levels of cyanide are variable (0 – 32mg HCN equivalent Kg^{-1}) depending on the processes, method and locality [5,6]. "Gari" has been shown to be a rich source of energy but of poor protein content (1.03%) compared with soy bean (44.08%) [7]. It has low levels of methionine, tryptophan, lysine and phenylalanine [8]. Supplementary protein sources must therefore be provided if cassava is to maintain its role as a major source of calories [4]. Many attempts have been made to enrich cassava products with protein from vegetable sources [9,10]. Oshodi [9] enriched "gari" with combination of soy grits and defatted melon. The protein content was increased from 1.43 to 19.41% dry basis with 40% protein supplement and 60% "gari". However the color and the odor were scored lower than the control. The 'eba' made from the enriched "gari" was darker in color. Collins and Temalilwa [11] increased the Protein Efficiency Ratio (PER) of cassava flour to 1.55 by adding to it 20% soy flour. Sanni and Sobamiwa [10] enriched "gari", after grating, with soybean residue after dewatering and full fat soy flour at 25% before and after fermentation. The protein content was raised from 9 to 11% resulting in a more nutritive and safer "gari". Banjo and Ikenebomeh [12] compared three stages of enriching "gari" with soy protein at the point of grating, before frying and after frying in varying "gari": soy flour ratios. The workers found that ratios 1:1 and 3:2 produced unacceptable soy enriched "gari" while the ratios 7:3 and 4:1 were acceptable The International Institute of Tropical Agriculture [13] recommended the use of soybean residual to fortify "gari" in a post-fermentation operation and found out that the taste of the soybean fortified "gari" was not different from that of the traditional "gari". Numfor and Noubi [14] studied the effect of full-fat soybean flour on the quality and acceptability of fermented cassava flour. While trying to improve the protein quality level of "gari", it must not be done in such a way as to affect the physicochemical and sensory properties. Oyewole and Asagbra [4] observed that co-fermentation of cassava with 20% cowpea and soy bean was found to increase the protein contents of the fermented cassava product from 1.8% to 5.5% and 8.2% respectively without affecting the organoleptic properties of the product. Past efforts have shown that using soybean, melon or groundnut alone was not sufficient in providing the necessary essential amino acids comparable to the reference protein. The reasons being that adequate amount of these flours cannot be added without strikingly altering the flavors, palatability and appearance of the "gari" product. Also, their biological values are not high enough to compensate for the small amounts in which they have to be added to "gari" and moreover they require further enrichment with Lysine and Methionine [15]. In an attempt to cut down on the amount of legumes used for enriching "gari" at the same time improving the amino acid content of fortified "gari". Oshodi [9] enriched "gari" with soy grits and defatted melon flour at 25% replacement level, separately and 40% level collectively. The limiting amino acids Lysine and Histidine in melon were provided by soybean while the Methionine lacking in soybean was supplied by melon. The protein content was increased from 1.43 to 19.4% in the enriched sample. The swelling capacity in cold water was retained but the color of the 'eba' was darker and the texture was rejected as too soft and lacked cohesiveness. The objective of this current study was to supplement cassava mash with different types of soy-melon protein supplement such as the defatted, full fat and residue after milk extraction from soy-melon milk in order to increase the nutritional and physicochemical qualities of "gari" and also determine the best type of supplement in terms of nutritional and physicochemical qualities.

2. MATERIALS AND METHODS

2.1 Source of Materials

Freshly harvested cassava roots were obtained from the research farm of the Federal University of Technology, Akure, Ondo State, Nigeria. Soybean and melon seeds used to produce the protein supplements were purchased from the Oba market, in Akure, Ondo State, Nigeria. They were sorted, cleaned, packed and kept under refrigeration until use.

2.2 Sample Preparation

2.2.1 Full fat soy flour

This was developed according to the methods of Sanni and Sobamiwa [10].

2.2.2 Melon flour

1 kg of melon seeds (Citrulis vulgaris) were toasted in an open pan over fire until light rcwn in color. They were milled in a Moulinex blender (1 single blade, Super Intermet, Japan) to a particle size of 450µm.

2.2.3 Defatted Soy-melon flour

Part of the milled soy and melon flours were defatted separately at room temperature with N-hexane until the residual oil was about 1.5%. These were used to supplement the grated cassava semolina before fermentation, after fermentation and after toasting.

2.2.4 Soy-melon milk residue

The sorted and cleaned soybean and melon seeds were boiled in water at 100ºC for about 25 minutes. The boiled soybean seeds were dehulled manually and both were wet milled separately in a hammer mill. Water was added in ratio 1:8 and a muslin cloth was used to extract the milk. The residues obtained were oven dried at 65ºC for 24 hrs [16]. The flours obtained were similarly used to supplement the cassava semolina before fermentation, after fermentation and after toasting.

2.2.5 Soy-melon "gari"

This was produced according to the methods of Banjo and Ikenebomeh [12]. The Cassava tubers were peeled manually with a sharp knife, washed and grated in a locally fabricated mechanical grater (Fig. 1).

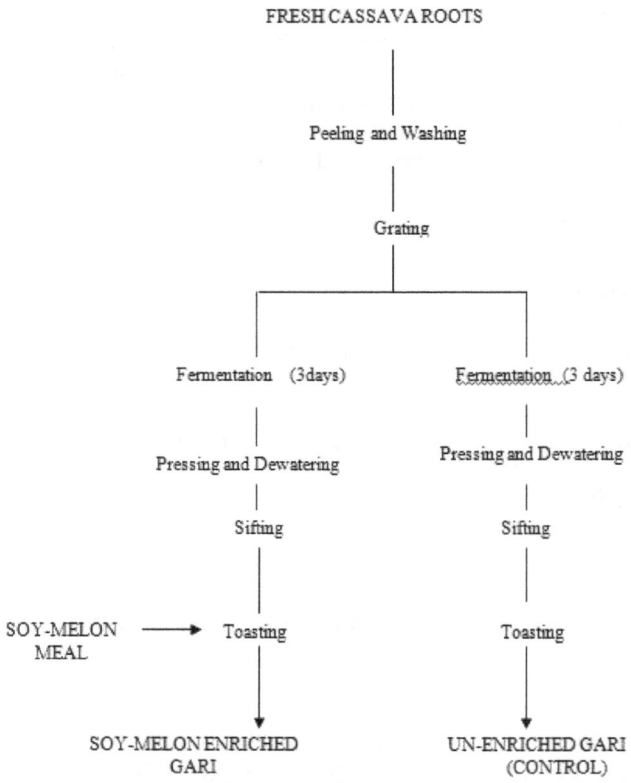

Fig. 1. Flow chart for the processing of soy-melon enriched and control gari.

The grater was made of a flat galvanized sheet punctured with holes with a big nail with opening of 0.75cm diameter and fixed round a drum-like plank. This was connected through a belt to a 7 hp driving motor [17]. The grated wet mash was then allowed to ferment for 72 hours after which it was dewatered in a mechanical press(Addis Engineering Nig. Ltd, Nigeria). The dewatered wet cassava cakes were pulverized with hands and sifted on a local raffia made sieve of mesh (0.3cm x 0.3cm) mounted on a rectangular wooden frame 40cm^2 to remove the fibers. The sifted meal was divided into four portions. Different types of soy-melon supplements (i) full fat soy-melon (ii) defatted soy-melon and (iii) soy-melon residue; were used to enrich the cassava meal during toasting (soak mix method) using 15% enrichment level and taking into consideration the water content of the mash of 65% [18]. The remaining batch was used as the control, containing no supplement. The white and fluffy meal was introduced into a wide aluminum pan (garifier) supported and being heated over wood fire. It was continuously stirred using a self insulating manual baffle made of calabash from gourd. This operation fairly distributes the heat to prevent or limit dextrization of "gari". The wet semolina fluffy meal was introduced into the garifier piecemeal, amidst continuous stirring, until a full manageable batch is subjectively determined as done. The time taken to get the batch toasted to dryness depends on the experience of the processor. From past studies, 1.5-2.2 kg of "gari" semolina was satisfactorily toasted over fire between 13-20 minutes, with high intensity of heat [17]. The toasted "gari" was removed from the wide aluminum pan, spread over a large spread of clean surface of woven Hessian sack and allowed to cool. The cooled "gari" samples were then packaged in High Density Polyethylene (HDPE) film and kept under refrigerated storage until ready for further analysis.

2.3 Analysis

2.3.1 Chemical analysis

The proximate compositions were determined according to the standard methods of AOAC [19]. The crude protein was determined by multiplying the total nitrogen by 6.25. The carbohydrate content was obtained by difference. The pH was measured with a pH meter. The total cyanide (mg/100g) was determined by the method of Rao and Hahn [20]. The phytic acid was determined by the method of Wheeler and Ferrel [21].

2.3.2 Physical analysis

The Bulk density was determined by the method of AOAC [19]. The density calculated here was referred to as Loosed density. The same container was used to determine the packed density after compacting by tapping the cylinder gently unto the wooden surface by dropping it form a height of 1.2cm once per second, adding more flour until the cylinder was full and the top was scrapped off with a spatula to obtain uniform volumes. The Swelling capacity was determined by the method of Ukpabi and Ndimele [22]. The Reconstitution Index was determined by the method of Banigo and Akpapunam [23]. The Wettability was determined by the method of Armstrong et al. [24].

2.3.3 Statistical analysis

Means and standard errors of the mean (SEM) of replicate scores were determined and subjected to analysis of variance (ANOVA) using the Statistical Package for Social Statistics (SPSS version 16). Means were separated using The Duncan's New Multiple Range (DNMR) Test [25].

3. RESULTS AND DISCUSSION

3.1 Chemical Composition of Soy-Melon Enriched "Gari"

3.1.1 Protein content

Enrichment significantly increased the protein content from 2.81% in the control sample to 15.27% for "gari" enriched with soy-melon milk residue; 19.92% for "gari" with full fat supplement and 23.54% for "gari" with defatted supplement (Table 1). This means that the defatted samples had the highest mean protein content followed by those with full fat supplements while those with soy melon milk residue were the lowest (p < 0.05). The highest values for the "gari" with defatted supplements might have been due to the removal of the oil in the supplement thus increasing the values of the protein contents [26]. Addition of soy meal into low protein foods such as sweet potato meal has been known to increase the protein content of such foods [27]. According to the recommendations of the Protein Advisory Group (PAG) guidelines [28] for weaning foods, protein content should be at least 20% (on a dry weight basis), fat levels up to 10%, moisture 5% to 10%, and total ash not more than 5%. The results for the composition of the enriched samples fall within these acceptable ranges of recommendations. The increase in protein content is attributable to the incorporation of soy meal in the blend. Therefore enrichment of "gari" with soy-melon meal produced "gari" of higher nutritional value. The protein contents of the blends (15.27 – 23.54%) were higher than the range of 11-14% recommended for growth by Beaton and Swiss [29] and still within the range of Protein Advisory Group (PAG) [28] guidelines. Therefore, the blends of soy-melon "gari" may be inferred as capable of supporting growth in school age children who often consume "gari" as a convenience food. To meet the Recommended Daily Allowance (RDA) of 34g/day for protein for a child of school age (11-14 years) consuming gari as a staple diet; he will need to consume about 1.209kg of normal control gari per day but with this protein enrichment he will need to consume less amount of about 0.222kg for gari enriched with soy melon residue, 0.170kg for gari enriched with full fat supplement and 0.144kg for gari enriched with defatted soy-melon supplement.

Table 1. Chemical composition of soy-melon enriched and un-enriched (control) "gari" samples

Parameter	Control	Defatted	Fullfat	Residue
Protein(%db)	$2.81^d\pm0.01$	$23.5^a\pm0.02$	$19.9^b\pm0.01$	$15.3^c\pm0.01$
Ash (%db)	$1.18^d\pm0.01$	$3.47^a\pm0.01$	$2.36^b\pm0.01$	$1.96^c\pm0.01$
Fat (%db)	$3.24^c\pm0.01$	$4.13^c\pm0.01$	$13.5^a\pm0.02$	$7.02^b\pm0.01$
Crude fiber (%db)	$6.31^a\pm0.01$	$4.79^d\pm0.01$	$5.92^b\pm0.01$	$5.16^c\pm0.01$
Carbohydrate(%db)	$86.5^a\pm0.01$	$64.1^c\pm0.67$	$58.3^d\pm0.01$	$70.6^b\pm0.01$
Total Energy (MJ/g)	$1.62^c\pm0.03$	$1.62^c\pm0.09$	$1.82^a\pm0.04$	$1.70^b\pm0.02$
pH	$3.62^c\pm0.02$	$5.25^a\pm0.01$	$4.94^b\pm0.01$	$4.86^b\pm0.01$
HCN (mg/kg)	$13.4^a\pm0.07$	$6.72^b\pm0.01$	$6.72^b\pm0.00$	$6.70^b\pm0.09$
Acidity (%)	$0.465^a\pm0.060$	$0.363^c\pm0.010$	$0.413^b\pm0.010$	$0.384^c\pm0.090$
Phytate (mg/kg)	$16.9^d\pm0.02$	$22.4^a\pm0.15$	$22.0^a\pm0.41$	$19.2^c\pm0.31$

Values are means of triplicate determinations.
Means in the same row bearing different superscript are significantly different. P < 0.05.
± are Standard error of Mean

3.1.2 Fat content

The fat content increased from 3.24% in the control sample to 4.13% in "gari" with defatted supplement, 7.02% in "gari" with milk residue and 13.50% in "gari" with full fat supplement. Expectedly, full fat enriched "gari" samples had the highest fat contents, followed by those enriched with milk residue and finally by "gari" enriched with defatted supplements. This increase in the fat content in "gari" enriched with full fat supplement might have been as a result of the contribution of the oil in the full fat soy-melon supplement [30].

Table 2. Physical properties of soy-melon enriched and un-enriched (control) "gari" samples.

Parameter	Control	Defatted	Fullfat	Residue
Swelling Index(v/v)	$4.80^a \pm 0.05$	$4.02^b \pm 0.04$	$4.01^b \pm 0.03$	$4.00^b \pm 0.05$
Reconstitution Index (v/v)	$4.50^a \pm 0.04$	$4.30^b \pm 0.09$	$4.35^b \pm 0.03$	$4.23^c \pm 0.03$
Wettability (seconds)	$30^c \pm 0$	$120^b \pm 2$	$120^b \pm 1$	$150^a \pm 1$
Packed Bulk Density (g/dm^3)	$0.610^b \pm 0.031$	$0.670^a \pm 0.021$	$0.660^a \pm 0.023$	$0.620^c \pm 0.031$
Loosed Bulk Density (g/dm^3)	$0.530^b \pm 0.022$	$0.580^a \pm 0.022$	$0.520^c \pm 0.032$	$0.500^d \pm 0.021$

Values are means of triplicate determinations.
Means in the same row bearing different superscript are significantly different. P < 0.05.
± are Standard error of Mean

3.1.3 Ash content

Enrichment with soy-melon meals resulted in higher ash contents. Enrichment increased the ash content significantly (P < 0.05) from 1.18% in the control "gari" to a minimum of 1.96% in "gari" enriched with residue from soy-melon milk to a maximum of 3.47% in "gari" enriched with defatted meal. This shows that the ash contents of "gari" with defatted residues were higher than those of "gari" enriched with either milk residue or full fat supplements. This is similar to the findings of Edem et al. [31], who increased the ash content of "gari" to 5.17% and 5.58% by fortifying with 10% and 15% soy meal respectively. This is also true of some food products other than "gari". Iwe and Onadipe [32] increased the ash contents of sweet potato from 2.2% to 2.5% - 4% by supplementing it with soy meal up to 25% level.

3.1.4 Crude fibre

Crude fibre decreased from a value of 6.31% to 4.79%, 5.16% and 5.92% for "gari" with defatted supplement, milk residue and full fat supplement respectively. This showed that enrichment with soy-melon supplement reduced the fibre content of "gari". This reduction of the fibre content might have been due to the dilution effect of the supplement on the fibre content of "gari".

3.1.5 Carbohydrate content

The carbohydrate content was reduced from 86.5% (db) to a minimum value of 58.3% in the sample enriched with full fat soy-melon meal, 64.1% in sample with defatted supplement and the maximum value of 70.6%(db) in the sample enriched with soy-melon milk residue. Similar decrease with increase in protein enrichment was reported in "Ugali", a Kenyan soy-enriched maize meal [30], soy-enriched rice [34], soy-sweet potato meal mixtures [27] and soy-sweet potato meal cookie [35].

3.1.6 Total energy value

The energy values increased significantly (P < 0.0.5) from 1.62MJ/g in the control sample to a maximum value of 1.82MJ/g in the sample enriched with full fat meal. The increase and the higher values of energy in the "gari" with full fat supplement might have been due to its oil content since oil has twice the energy for the same quantities for both protein and carbohydrate [36]. It could be inferred that "gari" enriched with defatted soy-melon meal was the richest in terms of the protein and ash contents, while "gari" enriched with full fat supplement was highest in fat content and total energy values while "gari" enriched with milk residue was high in crude fibre. Un-enriched "gari" (control) was highest in crude fibre and carbohydrate contents. These shows that "gari" enriched with defatted meal seem to be the best sample in terms of the proximate composition.

3.1.7 Hydrocyanic acid (HCN)

The hydrocyanic acid content was consistently lower in all the enriched samples than the control "gari". HCN was reduced from 13.5mg/kg in the control "gari" to a range of 6.72mg/kg – 12.5mg/kg for sample toasted together with full fat soy-melon meal. These values were lower than the recommended standard value of 20.0mg/kg [37,38,39]. They were also lower than a range of values reported in earlier studies [5,40,41]. This decrease in value might have been due to the dilution effect of the soybean protein in the supplements as observed by earlier workers [10,42]. Sanni [43] observed that after 3days of fermenting cassava, "gari" had a total cyanide content of 20.0 and 7.0mg/kg during the dry and wet seasons respectively. Sanni and Sobamiwa [10] also observed that the cyanide content of soy-"gari" was reduced with increasing protein content. The type of supplement added did not have any significant effect on the hydrocyanic acid content (P < 0.05).

3.1.8 pH/Acidity

The pH of "gari" was increased by enrichment with soy-melon meal. It increased from 3.62 in the control sample to the maximum value of 5.25 in the sample toasted together with defatted meal. The acidity was correspondingly reduced from 0.46% in the control "gari" to a minimum value of 0.36% lactic acid in the sample toasted together with defatted soy-melon meal. This showed that enrichment with soy meal tended to make the "gari" less acidic by the dilution effect of the supplements. This general increase in pH and the corresponding decrease in the acidity with enrichment might have been due to the dilution effect of the soy-melon supplement which indirectly responsible for the reduced sourness in the enriched "gari" samples [12]. This reduction by enrichment made the acidity to be much lower than the recommended range values of 0.9 - 1.2% [44] and 0.6% - 1.0% [38]. This was noticed in the reduction of the sour taste of the soy-melon "gari" which was an important parameter of sensory quality of "gari". This reduction might have been due to the differences in the number of days of fermentation and the dilution effect of the soy-melon meal supplements. This had also been observed to be due to the production of ammonia from the soy-melon protein degradation [12,45].

3.2 Physical Properties of Soy-Melon Enriched "Gari"

3.2.1 Swelling index

Data showed that the swelling index of the enriched "gari" was lower than that of the control "gari" (Table 2). The swelling index decreased with enrichment (p < 0.05). Swelling index was reduced from 4.80 (v/v) to a range of minimum value of 4.00 (v/v) in the sample enriched with milk residue to a maximum value of 4.02 (v/v) in sample enriched with defatted supplement. This is a confirmation of the findings of Banjo and Ikenebomeh [12] who also reported a reduced swelling with enrichment. It is also similar to the findings of Afoakwa, et al. [46] that swelling capacity decreased slightly with increasing soy meal concentration in "gari" from 10% to 30% enrichment level. The lower values in the enriched samples must have been due to the presence of lipids in the soy and melon flour which must have reduced the swelling capacities of the "gari" granules [47]. High swelling capacity has been shown to give a greater volume and more feeling of satiety per unit weight of "gari" [48].

3.2.2 Reconstitution Index

The reconstitution index was reduced with enrichment from 4.50 (v/v) in the control sample to a range of 4.23 (v/v) in the sample enriched with the milk residue to 4.35 (v/v) in sample enriched with full fat soy-melon supplement. These lower values of reconstitution index with the enriched samples must have been due to the presence of oil in the soy and melon flours. This behavior is similar to that of swelling index which was also found to decrease with soy flour enrichment. It has been shown that the presence of lipids acted as a buffer thus lowering the swelling power of starch granules [47].

3.2.3 Wettability

This was measured as the time taken in seconds by soy-melon "gari" granules to sink in water when dropped from a distance of 13cm from above the surface of the water. This is an indication whether "gari" will float as particles over the water or sink to the bottom [24]. The smaller the value in seconds the faster the ability of the "gari" to sink and the faster the better the "gari". The wettability was found to increase significantly from 30seconds in the control sample to a range of 120seconds in the defatted and full fat samples to 150seconds in the sample enriched with milk residue. High wettability values means that it will require more time for the enriched "gari" samples to sink in water and will float for more time on the surface of the cold water that the control sample. The samples enriched with milk residue had higher   wettability values than the rest of the samples. This is a bit surprising because one would have expected the sample with full fat supplement to behave this way. This might have been due to the processing method. The residue was not milled to fine particles after drying before being incorporated into the wet mass and the toasted "gari". This might have been responsible for the higher values for wettability. When this finding was compared with previous findings it was found that the wettability values ranging from 120 – 150 (seconds) for soy-melon enriched "gari" were higher than those of 27-35 (seconds) reported for un-enriched *D. alata yam* flour [49] and 42.5 (seconds) reported for un-enriched *D. rotundata* yam flour [50]. This was an indication that the un-enriched yam flours had similar wettability values with un-enriched "gari" but were denser and will sink faster than soy-melon enriched "gari" granules.

3.2.4 Bulk densities

The packed bulk densities increased generally with enrichment (P < 0.05%). The packed densities increased from 0.061g/cm^3 to values ranging from a minimum value of 0.620g/cm^3 in sample enriched with milk residue supplement to a maximum of 0.670g/cm^3 for the sample enriched with defatted supplement. Loose bulk density reduced slightly from 0.53g/dm^3 in the un-enriched sample to 0.50 and 0.52 in the sample enriched with full fat and milk residue respectively; while it increased to a maximum value of 0.580g/dm^3 in samples enriched with defatted supplement. Udensi et al. [49] observed that high bulk density increases the rate of dispersion of granules in water, which is important in the reconstitution of "gari" in hot water to produce reconstituted "gari" ('eba') dough. Udensi and Okaka [50] further observed that a slight increase in bulk density was paralleled by a progressive increase in wettability and water holding capacity. Brennan *et al.* (1976) also reported an increase in bulk density with an increase in the sinkability of powdered particles. From the above observations, it could be inferred that "gari" enriched with defatted supplement had the best wettability values, better swelling and reconstitution indices than "gari" enriched with full fat and milk residue supplements. The packed densities in all the samples were however consistently and significantly higher than the loosed bulk densities, which means that more quantity of "gari" can be packed for the enriched products than the control "gari" for the same specific volume [51].

4. CONCLUSION

From the results it could be concluded that enrichment with soy-melon protein supplement, improved the nutrient quality of "gari" especially the protein, fat and the ash contents. It also reduced the hydrocyanic acid content, thereby producing "gari" of higher quality and better safety. The acidity of the enriched samples was however reduced thus lowering the sourness of "gari". This may be an advantage for people who are not used to the sour taste of "gari". However, enrichment reduced the swelling and the reconstitution indices and also reduced the sink ability of "gari"; while it increased the packed bulk densities. Soy-melon "gari" enriched with defatted supplement had therefore been shown to have the highest protein and ash contents, lowest crude fat and acidity than other enriched gari samples. It had also been shown from this study to have better wettability, water holding capacity, and ability to disperse in water. It also had better swelling and reconstitution indices than "gari" enriched with full fat and milk residue supplements.

COMPETING INTERESTS

Authors have declared that no competing interests exist.

REFERENCES

1.   Lancaster PA, Ingram JS, Lim MY, Coursey DG. Traditional cassava based foods. Survey of processing techniques. Economic Botany. 1982;36:pp.12–45.
2.   Kordylas JM. Processing and preservation of tropical and sub-tropical foods. ,Macmillan, London and Basingstoke; 1990.
3.   Osho SM. The processing and acceptability of a fortified cassava-based product ("gari") with soybean. Nutrition & Food Science. 2003;33(6):278-283.

4. Oyewole OB, Asagbra Y. Improving traditional cassava processing for nutritional enhancement. International Workshop on Food-based approaches for a healthy nutrition. Ouagadougou. 2003:23-28 / 11 / @. Available: http://www.univ-ouaga.bf/fn2ouaga2003/abstracts/0602_FP_Int3_Nigeria_Oyewole.pdf.

5. Ojo O, Deane R. Effects of cassava processing methods on anti-nutritional components and health status of children. J. Sci. Food Agric. 2002;82:252-257.

6. Oke OL. Eliminating cyanogens from cassava through processing, technology and tradition. Acta Hort. 1994;375:163–174.

7. FAO. Agriculture, Food and Nutrition for Africa – a Resource Book for Teachers of Agriculture, Geneva. Switzerland; 1997.

8. Okigbo BN. Nutritional implications of projects giving high priority to the production of staples of low nutritive quality. The case of cassava in the humid tropics of West Africa. Food and Nutrition Bulletin, United Nation University, Tokyo. 1980;2(4):1-10.

9. Oshodi AA. Protein enrichment of foods that are protein deficient. Fortification of "ĉari" with soybean and melon. Nig. J. Appl. Sci. 1985;3(2):15–22.

10. Sanni MO, Sobamiwa AO. Processing and characteristics of soybean – fortified "gari". World Journal Micro. and Biotechnology. 1994;10:268–270.

11. Collins JL, Temalilwa CR. Cassava *(Manihot esculenta Crantz)* flour fortification with soy flour. J. Food Sci. 1981;46:1025–1028.

12. Banjo NO, Ikenebomeh MJ. Comparison of methods for the preparation of soy "gari" from cassava and soybean mash. J. Food Sci. & Tech. 1996;33(4):440–442.

13. ITA. Nigeria tops the world in cassava production. In Tropical root and tuber crops. Bulletin 6:8 Ibadan, IITA; 1992.

14. Numfor FA, Noubi L. Effect of full-fat soya bean flour on the quality and acceptability of fermented cassava flour. Food and Nutrition Bulletin. Volume 16, Number 3; 1995.

15. Akinrele IA. Nutrient enrichment of "gari". W. A. J. Biological & Applied Chem. 1967;10(1):19-23.

16. Anonymous. Soybean for good health. How to grow and use soybeans in Nigeria. Ibadan. International Institute of Tropical Agriculture; 1990.

17. Agunbiade SO. Effect of fermentation period on the physicochemical properties of "gari". Bioscience Research Communications. 2001;13(1):65-71.

18. Akingbala JO, Oguntimehin GB, Almazan AM. Effect of processing factors on the cyanide content and acceptability of "gari". J. Food Processing and Preservation. 1987;17:337-380.

19. AOAC. Official methods of Analysis (15th Ed). Association of Official Analytical Chemist, Washington, D.C. USA; 1990.

20. Rao PV, Hahn SK. An automatic assay for determining the cyanide content of cassava *(Manihot esculenta Crantz)* and cassava products. J. Sci. Food Agric. 1984;35:426–436.

21. Wheeler EL, Ferrel RE. A method for phytic acid determination in wheat flour. Cereal Chem. 1971;48:312–316.

22. Ukpabi UJ, Ndimele C. Evaluation of the quality of "gari" produced in Imo State. Nig. Food. J. 1990;8:105-110.

23. Banigo EB, Akpapunam MA. Physicochemical and nutritional evaluation of protein enriched fermented maize flour. Nig. Food J. 1987;5:30-36.

24. Armstrong BI, Sternly DW, Maurice TJ. In: Functionality of Protein Structure. Akwa Pour-El Acs symposium Series 92; 1979.

25. Steel R, Torrie J, Dickey D. Principles and Procedures of Statistics: A Biometrical Approach", 3rd ed., McGraw Hill Book Co., New York, USA; 1997.

26. Iwe MO, Onuh JO. Functional and sensory properties of soybean and sweet potato flour mixtures. Lebesm-Wiss. U. Technol. 1992;25:569-573.

27. Iwe MO, Ngoddy PO. Proximate composition and some functional properties of extrusion cooked soybean and sweet potato blends. Plant Foods for Human Nutrition. 1998;53:121-132

28. PAG. Protein Advisory Group of the United Nations, PAG guideline no. 8, Protein-rich mixtures for use as weaning foods. New York: FAO/WHO/UNICEF; 1971.

29. Beaton H, Swiss LD. Estimation of protein energy ratio. Ame. J. Clin. Nutrition. 1974;27:485-489.

30. Nyotu HG, Alli I, Paquette G. Soy enrichment of a maize based Kenyan food (Ugali). J. of Food Sci. 1996;51(5):1204-1206.

31. Edem DO, Ayatse JO I, Itam E H. Effect of soy protein enrichment on the nutritive value of "'gari'" (farina) from Manihot esculenta. Food Chemistry. 2001;75:57-62

32. Iwe MO, Onadipe OO. Effect of addition of extruded full fat soy flour into sweet potato flour on functional properties of the mixture. J. Sustain. Agric. Environ. 2001;3(1):109-117.

33. Chauhan GS, Bains GS. The storage studies on full fat soy flour. Indian J. Nutrition and Dietetics. 1985;22(4):121-124.

34. Iwe MO. Proximate, physical and sensory properties of soy-sweet potato flour cookie. Global J. Pure and Applied Sciences. 2002;8(2):187-192.

35. Ingram JS. Standard specifications and quality requirements for processed cassava products. Tropical product Institute, London. 1975;pp11–2.

36. Osborne DR, Voogt P. Calculation of calorific value. In "The analysis of Nutrients in Foods, p. 239. Academic Press, London, New York; 1978.

37. CODEX STAN 151. Codex Alimentarius Commission. Codex STAN 151, African Regional Standard for "gari" (1989, Reviewed 1-1995)

38. NIS 181 (2004) Standard for "gari". In: Standard for cassava products and guidelines for export. Sanni et al., (2005). (Eds). IITA, Ibadan, Nigeria

39. Aletor VA. Cyanide in "gari". 1. Distribution of total, bound and free hydrocyanic acid in commercial "gari", and the effect of fermentation time on residual cyanide content. International Journal of Food Sciences and Nutrition. 1993;44:pp.281–287.

40. Adindu MN, Olayemi FF, Nze-Dike OU. Cyanogenic potential of some cassava products in Port Harcourt markets in Nigeria. Journal of Food Composition and Analysis. 2003;16:pp.21–24.

41. Oluwamukomi MO, Adeyemi IA, Oluwalana IB. Effects of soybean enrichment on the physicochemical and sensory properties of "gari". Applied Tropical Agriculture. Vol 10, Special issue. 2005;pp44-49.

42. Sanni LO. Delineating the quality criteria of "gari". Proceeding of a regional workshop on traditional African foods- Quality and Nutrition; 1991.

43. Edwards CC, Onyekwere OO, Akinrele IA. Preliminary studies on protein enrichment of "gari" with peanut grits: 1. Formulation, characterization and organoleptic evaluation. Proceedings of Nig. Inst. Food Sci. and Tech. 1977;1:95-102.

44. Reddy NR, Pierson MD, Salunkhe JOK. Idli. In: Legume-based fermented foods. CRC Press Inc. Boca Raton, Florida. 1986;pp145–160.

45. Afoakwa EO, Edem K, George A, Joseph A. Functional properties and sensory characteristics of soy-fortified "gari" Paper presented at the Second Pan- African Conference on IT in the Advancement of Nutrition in Africa (ITANA 2005). Cape Town, South Africa, September 18–21.Sep; 2005. Available at: http://works.bepress.com/georgeamponsahannor/9.

46. Cheftel JC. Chemical and nutritional modifications of food protein due to processing and storage. In: Food Proteins. (Eds). J. R. Whitaker and S. R. Tannenbaum. AVI Publ. Westport CT; 1977.

47.	Almazan AM. Influence of cassava variety and storage on "gari" quality. Trop. Agric (Trinidad). 1992;69(4):386–390.

48.	Udensi EA, Oselebe HO, Iweala OO. The Investigation of Chemical Composition and Functional Properties of Water Yam (*Dioscorea alata*): Effect of Varietal Differences. Pakistan Journal of Nutrition. 2008;7(2):342-344.

49.	Udensi EA, Okaka JC. Predicting the effect of particle size profile, blanching and drying temperature on the dispersibility of yam flour. Global J. Pure Appl. Sci. 2000;6: 589-592.

50.	Fagbemi TN. Effect of blanching and ripening on functional properties of Plantain (*Musa* aab) Flour. Plant Food for Human Nutrition. 1999;54:261–269.

12

Nutritional Quality of Cookies Produced From Mixtures of Fermented Pigeon Pea, Germinated Sorghum and Cocoyam Flours

L. C. Okpala[1*] and O. O. Ekwe[2]

[1]Department of Food Science and Technology, Ebonyi State University, Abakaliki, Nigeria.
[2]Department of Animal Science, Ebonyi State University, Abakaliki, Nigeria.

Authors' contributions

This work was carried out in collaboration between both authors. Author OOE designed the experimental animal housing, while author LCO managed the analyses of the study, conducted the literature searches and wrote the manuscript. Both authors read and approved the final manuscript.

ABSTRACT

Due to the high foreign exchange spent on the importation of wheat and the need to combat issues of malnutrition in developing countries, cookies were produced from flour blends of fermented pigeon pea (FPF), germinated sorghum (GSF) and cocoyam (CF). Proximate composition of the cookies revealed that cookies made with 100%FPF had the highest protein content of 16.13% while cookies made with 100%CF had the least protein value of 6.40%. The antinutritional factors investigated in the cookies were low and within allowable limits. The nutritional quality evaluated by animal feeding experiments revealed that biological values (BV) of cookies ranged from 78.16% (for 100%CF) to 96.57% (for 33.3%FPF:33.3%GSF:33.3%CF); net protein utilization (NPU) values ranged from 70.08% (for 100%CF) to 92.98% (for 33.3%FPF:33.3%GSF:33.3%CF) while true digestibility (TD) ranged from 89.53%(for 100%CF) to 97.88% (for 66.6%FPF:16.7%GSF:16.7%CF). The results obtained suggest that cookies of good nutritional value can be produced from these locally available crops.

Keywords: Cookies; nutritional quality; composite flours.

*Corresponding author: Email: lcokpala@yahoo.com

1. INTRODUCTION

Cookies are snacks which are widely consumed all over the world especially by children. Wheat flour which is used to produce cookies is unavailable in many regions of the world resulting in importation of the flour by regions with limited supplies. There is therefore, a compelling need to develop an adequate substitute for wheat flour. This substitute should be readily available, cheap and replace wheat in functionality. Composite flour can be defined as a mixture of several flours obtained from roots and tubers, cereals, legumes etc. with or without the addition of wheat flour [1]. Composite flours have been used extensively in the production of baked goods. In fact, several attempts have been made to produce cookies from different types of composite flours [2,3,4]. In countries where malnutrition poses a serious problem especially among children, composite flours which have better nutritional quality would be highly desirable.

Fermentation is an age long process of processing cereals and legumes which not only modifies some physical characteristics of these grains but also increases the levels of nutrients, digestibility and bioavailability as well as decreases the level of antinutrients [5]. Germination is another traditional method of processing grains which has resulted in increasing free limiting amino acids and available vitamins with modified functional properties of seed components [6]. With such potentials, the use of composite flours made from grains which have undergone fermentation and/or germination should result in products of enhanced nutritional quality.

Pigeon pea (*Cajanus cajan*) is an underutilized legume which grows in the tropics and sub-tropics of India, Africa, South-east Asia and Central America [7]. It has a protein content of about 25.83% [8]. Sorghum (*Sorghum bicolor*), a cereal is a member of the grass family. It is the fifth most important cereal in the world after wheat, rice, corn and barley. More than 35% of sorghum is grown for human consumption while the rest is primarily used for feed and industrial alcohol production [9]. Available data from FAO [10], reports that there is a decrease in consumption in parts of Africa and this has been attributed to shift in consumer habits brought about by rapid rate of urbanization, time and energy required to prepare food based on sorghum as well as poor marketing facilities and processing techniques. Cocoyam (*Xanthosoma sagittifolium*) is an edible root crop grown in the tropics of which Nigeria is a major producer. It belongs to the family Aracea [11]. The flour from cocoyam has been used in baking of products as it has been reported that cocoyam has fine granular starch which improves binding and reduces breakage of snack products [12].

All these crops are grown in large quantities but are grossly underutilized. This work therefore seeks to produce cookies from flour blends of fermented pigeon pea, germinated sorghum and cocoyam with the aim of producing products of high nutritional quality.

2. MATERIALS AND METHODS

2.1 Materials Used for Cookie Preparation

The white variety of pigeon pea (*Cajanus cajan*), the white variety of sorghum (*Sorghum bicolor*) and the tannia variety of cocoyam (*Xanthosoma sagittifolium)* were purchased from a retail outlet in Abakaliki, Ebonyi State, Nigeria. Wheat flour and all other baking ingredients such as eggs, baking powder, fat, milk and flavourings were also obtained from the same source.

2.2 Method of Preparation

2.2.1 Cocoyam flour (CF)

Cocoyam flour was produced using the method described by Udensi et al [13] with slight modifications. The corms were washed, peeled, sliced and blanched at 80°C for four minutes. They were dried and milled to pass through 100µm mesh sieve.

2.2.2 Fermented pigeon pea flour (FPF)

The pigeon pea flour was subjected to natural lactic acid fermentation using the method of Hallén et al [6]. Washed and dried grains were ground into fine flour. The flour was mixed with water (1:5 wt/wt) to form slurry followed by the addition of 5% sugar by weight of flour. The slurry was left to ferment in trays at room temperature until the pH of the slurry reached 5.5. The fermented slurry was dried at 50°C and then sieved through a 100µm mesh screen.

2.2.3 Germinated sorghum flour (GSF)

Sorghum grains were germinated using the modified method of Hallén et al [6]. Cleaned grains were soaked in 0.1% sodium hypochlorite solution for 30 minutes to prevent mould growth. The grains were thoroughly washed and soaked in water (24 hours). The hydrated seeds were spread on jute bags and allowed to germinate for 4 days after which they were dried at 50°C. Thereafter, formed roots and testa were rubbed off before milling and sieving through 100µm mesh sieve. All of the flour samples were kept in airtight containers until needed for analysis.

2.2.4 Formulation of flour composites

Ten formulations were generated (Table 1). The flours were thoroughly mixed to obtain homogeneous blends. Samples were stored in airtight containers at room temperature until ready for use.

Table 1. Formulation of composite flour

Formulation	Cocoyam (CF)	Fermented pigeon pea (FPF)	Germinated sorghum (GSF)
1	100	-	-
2	-	100	-
3	-	-	100
4	50	50	-
5	-	50	50
6	50	-	50
7	33.3	33.3	33.3
8	66.6	16.7	16.7
9	16.7	66.6	16.7
10	16.7	16.7	66.6

2.3 Cookie Preparation

The ingredients used were: flour, 100.0g; hydrogenated vegetable fat, 40.0g; sugar (granulated cane), 25.0g; egg (whole, fresh), 31.0g; milk (full-fat filled, powdered); 7.8g;

nutmeg; 0.3g, vanilla (liquid), 5.0ml, salt, 1.0g and baking powder, 1.0g. Fat and sugar were creamed using an electric mixer at medium speed for 5 minutes. Eggs and milk were added while mixing and then mixed for a total of about 30 minutes. Vanilla, nutmeg, flour, baking powder and salt were mixed thoroughly and added to the cream mixture where they were all mixed together to form a dough. The dough was rolled and cut into circular shapes of 5cm diameter. Baking was carried out at 185°C for 20±5 minutes. Cookie samples were cooled and stored in airtight containers until needed. Cookies were made from wheat to serve as a control.

2.4 Laboratory Analysis

2.4.1 Proximate composition of cookies

Protein content was determined using the micro – Kjeldahl method as described AOAC [14]. Fat, ash, fibre and moisture contents were also determined according to the methods described by AOAC [14]. Carbohydrates were determined by difference. Analyses were performed in triplicates

2.4.2 Evaluation of antinutritional factors

Phytic acid was determined using the method of Reddy and Love [15]. Saponins were determined using the method of Birk et al [16] as modified by Hudson and El- Difrawi [17], hydrogen cyanide was determined using the alkaline picrate spectrophotometric method as described by Balagopalan et al [18], oxalate was determined using the method described by Ukpabi and Ejidoh [19], tannins were determined using the method of Folin- Dennis [14], while trypsin inhibitor activity was determined using the spectrophotometric method described by Arntifield et al [20]. Analyses were performed in triplicates.

2.4.3 Animals and housing

Fifteen groups (Ten composite cookie formulations groups + one wheat cookie group + three casein control groups + one group receiving a protein-free diet) of five male adult albino rats of the Wistar strain with average initial weight of 120-210g were used. They were divided in such a way that all the groups of rats had the same average weight. Subsequently, they were housed in individual screened bottomed cages designed to separately collect faeces and urine. Experimental animals each received 20g of the corresponding group diets and water *ad libitum*. The temperature of the laboratory was 28 ± 1°C with alternate 12h periods of light and dark. These animals were used to assess the BV and NPU of the diets based on casein. Following the method described by Al- Numair and Ahmed [21], a 9-day balance study which included a four-day adjustment and five-day nitrogen (N) balance period was carried out. There was a preliminary feeding period of four days followed by a balance period of five days during which complete collection of faeces and urine was performed for each rat. Food intake was monitored daily and final body weights were recorded. Urine was collected in sample bottles, preserved in 0.1N HCl to prevent loss of ammonia and stored in a refrigerator until analyzed for urinary nitrogen. Faeces of individual rats were pooled, dried at 85°C for 4 hours, weighed before being ground into fine powder and stored for faecal N determination. The concentration of nitrogen in the diet, faeces and urine was estimated by the Kjeldahl method [14].

2.4.4 Diets

Formulations of the test diets are shown in Table 2. The diets were adjusted to provide 6, 8 and 10% protein. The diets made from cookies with protein content above 10% were adjusted to provide 10% protein; test diets made from cookies with protein content ranging from 8.1-9.9% were adjusted to provide 8% protein while diets from cookies with protein content ranging from 6.1-7.9% were adjusted to provide 6% protein. There were three reference protein diets of casein at 6, 8 and 10% protein levels for comparison of the protein quality of these test diets [22]. Other ingredients included 5% vegetable oil, 0.25% vitamins, 0.05% minerals and the remainder, corn starch added to balance the diets [23]. The diets were thoroughly mixed, pelletized and stored in polyethylene bags labeled with designated names. The polyethylene bags were kept in airtight containers until ready for use.

2.5 Statistical Analysis

The data collected were subjected to analysis of variance (ANOVA). Means were separated using Duncan's multiple range test using the Statistical Package for the Social Sciences (SPSS) version 16.0 (SPSS Inc., Chicago, IL. USA).

3. RESULTS AND DISCUSSION

3.1 Proximate Composition of Cookies

The chemical composition of cookies produced is presented in Table 3. The protein content of the cookies ranged from 6.40 to 16.13%. Cookies made from 100% fermented pigeon pea flour (100FPF) had the highest protein content and it was significantly higher ($p < 0.05$) than all the other cookies. This was closely followed by cookies made from 100% germinated sorghum flour (100GSF). It was observed that increase in FPF resulted in an increase in the protein content of the cookies. The ash contents were observed to increase with increase in cocoyam flour (CF). Ash content is indicative of the amount of minerals in any food sample [24]. The fat content of the cookies were relatively low. Fat plays a significant role in the shelf life of food products as it can promote rancidity in foods, leading to the development of unpleasant and odorous compounds [11]. As such, relatively low fat content is desirable in baked food products.

3.1.1 Antinutritional factors

The levels of antinutritional factors in the cookies produced (Table 4) were much lower than the values obtained by Okpala and Okoli [25] for biscuits made with pigeon pea, sorghum and cocoyam flour blends which did not undergo any form of fermentation or germination. This suggests that germination of sorghum and fermentation of pigeon pea led to reduction in the antinutritional factors present in the flours. All the values obtained were low and within safe limits.

Table 2. Formulation of fermented pigeon pea, germinated sorghum and blanched cocoyam flour cookies and casein based diets fed to rats

Ingredients	Protein (%)	100GSF	100FPF	50GSF:50FPF	66.6FPF:16.7CF:16.7GSF	Casein A	50CF:50GSF	50CF:50FPF
100GSF	10	648.5	-	-	-	-	-	-
100FPF	10	-	620.4	-	-	-	-	-
50GSF:50FPF	10	-	-	673.0	-	-	-	-
66.6FPF:16.7CF:16.7GSF	10	-	-	-	629.7	-	-	-
Casein A	10	-	-	-	-	110.1	-	-
50CF:50GSF	8	-	-	-	-	-	813.0	-
50CF:50FPF	8	-	-	-	-	-	-	892.9
Oil	-	50	50	50	50	50	50	50
Vitamins	-	2.5	2.5	2.5	2.5	2.5	2.5	2.5
Minerals	-	0.5	0.5	0.5	0.5	0.5	0.5	0.5
Corn starch	-	298.5	326.6	274.0	317.3	836.9	134.0	54.1
Total	-	1000	1000	1000	1000	1000	1000	1000

Ingredients	Protein (%)	33.3CF:33.3GSF:33.3FPF	66.6GSF:16.7CF:16.7FPF	66.6CF:16.7GSF:16.7FPF	100 WHEAT	Casein B	100CF	Casein C
33.3CF:33.3GSF:33.3FPF	8	806.50	-	-	-	-	-	-
66.6GSF:16.7CF:16.7FPF	8	-	809.7	-	-	-	-	-
66.6CF:16.7GSF:16.7FPF	8	-	-	801.6	-	-	-	-
100 WHEAT	8	-	-	-	829.0	-	-	-
Casein B	8	-	-	-	-	88.1	-	-
100CF	6	-	-	-	-	-	872.1	-
Casein C	6	-	-	-	-	-	-	66.1
Oil	-	50	50	50	50	50	50	50
Vitamins	-	2.5	2.5	2.5	2.5	2.5	2.5	2.5
Minerals	-	0.5	0.5	0.5	0.5	0.5	0.5	0.5
Corn starch	-	140.5	137.3	145.4	118.0	858.9	74.9	880.9
Total	-	1000	1000	1000	1000	1000	1000	1000

CF= Cocoyam flour; GSF= Germinated sorghum flour; FPF= Fermented pigeon pea

Table 3. Chemical composition of cookies produced from fermented pigeon pea, germinated sorghum and blanched cocoyam flour blends

Blends CF:GSF:FPF	Moisture (%)	Fat (%)	Protein (%)	Ash (%)	Fibre (%)	Carbohydrates (%)	Energy (kcal/100g)
100:0:0	6.85^i	5.10^k	6.40^k	2.95^a	2.26^{de}	76.44^a	377.35^b
0:100:0	8.88^c	5.48^j	15.42^c	2.17^f	2.33^c	65.69^g	373.76^c
0:0:100	8.85^d	6.32^a	16.13^a	2.36^c	2.23^e	64.12^j	377.88^b
50:50:0	8.47^f	5.95^e	9.84^h	2.35^{cd}	2.29^{cd}	71.10^d	377.31^b
0:50:50	8.46^f	6.12^c	14.86^d	2.05^g	2.28^d	66.23^f	379.44^a
50:0:50	8.12^g	6.05^d	8.96^j	2.18^f	2.13^f	72.56^b	380.53^a
33.3:33.3:33.3	8.62^f	5.79^h	9.92^f	2.28^{de}	2.33^{bc}	71.06^d	376.03^b
16.7:16.7:66.6	9.02^a	6.22^b	15.88^b	2.36^c	2.08^f	64.37^h	376.98^b
16.7:66.6:16.7	8.94^b	5.83^g	9.88^g	2.26^e	2.37^{ab}	70.72^e	374.87^{bc}
66.6:16.7:16.7	8.77^e	5.87^f	9.98^e	2.42^c	2.25^{de}	70.72^e	375.63^b
100% wheat flour	7.46^h	5.64^i	9.65^i	2.56^b	2.41^a	72.28^c	378.48^{ab}

CF= Cocoyam flour; GSF= Germinated sorghum flour; FPF= Fermented pigeon pea
Means within a column with the same superscript are not significantly different (p> 0.05).
Values in Tables 3-4 are means of triplicate determinations.

Table 4. Antinutritional factors of cookies produced from fermented pigeon pea, germinated sorghum and blanched cocoyam flour blends

Blends CF:GSF:FPF	Saponins (mg/100g)	Hydrogen Cyanide (mg/kg)	Phytic acid (mg/100g)	Tannins (mg/100g)	Oxalate (mg/100g)	Trypsin Inhibitor (TIU/mg)
100:0:0	0.04^a	0.06^b	0.70^a	0.46^{ab}	5.70^a	0.04^b
0:100:0	0.03^a	0.11^a	0.62^c	0.32^c	4.00^c	0.06^a
0:0:100	0.04^a	0.11^a	0.55^e	0.37^{bc}	4.20^c	0.06^a
50:50:0	0.03^a	0.10^{ab}	0.63^{bc}	0.37^{bc}	3.30^e	0.06^a
0:50:50	0.03^a	0.09^{bc}	0.58^d	0.39^{abc}	3.40^e	0.05^{ab}
50:0:50	0.03^a	0.08^{cd}	0.58^d	0.38^{abc}	3.70^d	0.05^{ab}
33.3:33.3:33.3	0.03^a	0.07^d	0.53^f	0.34^c	3.70^d	0.05^{ab}
16.7:16.7:66.6	0.03^a	0.08^{cd}	0.64^b	0.39^{abc}	3.60^d	0.04^b
16.7:66.6:16.7	0.04^a	0.10^{ab}	0.62^c	0.37^{bc}	3.10^f	0.05^{ab}
66.6:16.7:16.7	0.03^a	0.10^{ab}	0.47^g	0.45^{ab}	3.40^e	0.06^a
100% wheat flour	0.04^a	0.06^b	0.69^a	0.47^a	5.50^b	0.05^{ab}

CF= Cocoyam flour; GSF= Germinated sorghum flour; FPF= Fermented pigeon pea
Means within a column with the same superscript are not significantly different (p> 0.05).

3.1.2 Nutritional evaluation

Table 5 shows the nutritional evaluation of experimental diets. It was observed that there were no significant differences (p>0.05) in the food intake of rats fed the casein based diets at all the levels of protein. However for the test diets which provided 10% protein, the food intake of the diet formulated with 100FPF cookies was significantly lower (p<0.05) than all other diets in the same group with the exception of that of 16.7CF:16.7GSF:66.6FPF. The reduced food intakes could be as a result of poor palatability of the diets. Rats fed diets from 100CF (which provided 6% protein) had the least food intake. Mercer et al, [26] demonstrated that the dietary concentration of protein or amino acids regulates many aspects of food intake, growth and growth efficiency of rats. They further reported that at very low or very high concentrations of protein in the diet, there is a marked depression in food intake by rats. De Angelis et al, [27] also demonstrated a relationship between the ratio of non-essential amino acids to essential amino acids in a protein and that proteins ability to stimulate food intake. These findings therefore suggest that the low food intake by rats fed diets from 100CF could actually be as a result of the low protein concentration and/or possibly the essential amino acid profile.

With the exception of diets from 10% casein and 50GSF:50FPF, the nitrogen (N) intake of the 100GSF group was significantly higher (p<0.05) than all the experimental diets. There was no significant difference (p>0.05) in the N intakes of the 10% casein and 50GSF:50FPF groups. The high N intakes could be due could be due to the high food intake by the animals. Nnam [28] observed that high food intake was associated with high mean nitrogen intake.

Diets with high digested N but low urinary N resulted in high nitrogen retention, biological values and net protein utilization as seen with wheat cookies. Similar findings were reported by Onweluzo and Nwabugwu [29]. In the same vein, those diets with low digested N but high urinary N resulted in low nitrogen retention, biological values and net protein utilization as seen with diets from 100CF. It has been reported that lower values of N balance are indicative of lower protein quality [30].

With the exception of test diets made from 100CF, all of the diets compared favourably with the casein diets with respect to protein quality. Some of these test diets even exhibited superior protein quality than some of the casein diets. The rats fed the diet made from 33.3CF:33.3GSF:33.3FPF had the highest biological value (BV) of 96.57%. About four other groups of rats (fed other cookie based diets) had BVs that were not significantly different (p>0.05) from these values. The biological values of all the diets were higher than the recommended value of 75% for children [31]. The net protein utilization (NPU) values were also observed to be within the values (>70%) recommended for good food protein and dietary mixture [32]. True digestibility (TD) values were also high and this could be due to the reduction in antinutritional factors which are often responsible for the poor digestibility of proteins [33]. The cookies were generally of good protein quality with high values of BV, NPU and TD.

Table 5. Food Intake and Nitrogen balance of rats fed diets of casein and cookies produced from fermented pigeon pea, germinated sorghum and blanched cocoyam flour blends

Diets	% Protein	Food intake (g)	Nitrogen intake (g)	Fecal nitrogen (g)	Digested nitrogen (g)	Urinary nitrogen (g)	Nitrogen balance (g)	BV (%)	NPU (%)	TD (%)
100FPF	10	70.26a	1.12a	0.09b	1.03a	0.11a	0.92a	88.91e	84.12cde	94.62de
50GSF:50FPF	10	49.92b	0.80bc	0.06c	0.74cdef	0.12a	0.61cd	82.58f	79.00e	95.64cd
66.6FPF:16.7CF:16.7GSF	10	63.44a	1.01ab	0.07c	0.94ab	0.11a	0.84a	89.02de	85.90bcd	96.47bcd
Casein A	10	60.85ab	0.97ab	0.05cde	0.92ab	0.07bc	0.86a	93.14abcde	91.13ab	97.88abc
50CF:50GSF	10	68.03a	1.09a	0.14a	0.95ab	0.10ab	0.85a	89.84de	83.25de	93.63e
50CF:50FPF	8	64.28bc	0.82bc	0.07c	0.75bcde	0.03ef	0.72bc	95.78ab	89.90ab	93.86e
33.3CF:33.3GSF:33.3FPF	8	45.20c	0.58c	0.04cde	0.54f	0.02f	0.52d	94.99abc	91.03ab	95.85cd
66.6GSF:16.7CF:16.7FPF	8	67.32a	0.86bc	0.06cd	0.80bcd	0.03def	0.78abc	96.57a	92.98a	96.27bcd
66.6CF:16.7GSF:16.7FPF	8	47.91b	0.61c	0.05cde	0.56ef	0.06cd	0.51d	90.08abcd	86.44bcd	95.62cd
100 WHEAT	8	48.61b	0.63bc	0.04def	0.59ef	0.03def	0.56d	94.51abcd	92.60a	97.98abc
Casein B	8	65.86a	0.84bc	0.04def	0.80bcd	0.04def	0.77abc	95.13abc	93.63a	98.11abc
	8	60.65bcd	0.78bcd	0.02f	0.76bcde	0.06cd	0.70bc	92.80abcde	92.02ab	99.32a
100CF	6	26.30c	0.26d	0.05cde	0.21g	0.04def	0.17f	78.16g	70.08f	89.53f
Casein C	6	68.10c	0.66c	0.03c	0.63def	0.06cd	0.57d	90.68bcde	89.64abc	98.83ab

CF= Cocoyam flour; GSF= Germinated Sorghum flour; FPF= Fermented Pigeon pea flour
Means within a column with the same superscript are not significantly different (p> 0.05).
BV=Biological Value; NPU=Net Protein Utilization; TD=True Digestibility
Values are means obtained from 5 rats.

4. CONCLUSION

This study has shown that fermentation and germination of grains generally improved the nutritional quality of cookies produced from them. The use of fermentation and germination of the grains not only reduced the antinutritional factors but also improved the protein quality of the cookies. This study has also opened new possibilities of the application of flours produced from grains used in this study. Cookies produced from such composites will not only increase savings in foreign exchange for countries that rely heavily on the importation of wheat, but will also improve utilization of locally available crops and lead to enhanced nutrient intake by the consumer.

COMPETING INTERESTS

Authors have declared that no competing interests exist.

REFERENCES

1. Adeyemi SAO, Ogazi PO. The place of plantain in composite flour. Commerce Industry. Lagos State, Nigeria. World Health Organization (WHO) Rep. ser. 1973 No. 522 Committee. WHO Geneva; 1985.
2. McWatters KH, Ouedraogo JB, Resurreccion AVA, Hung Y, Phillips RD. Physical and sensory characteristics of sugar cookies containing mixtures of wheat, fonio (*Digitaria exilis*) and cowpea (*Vigna unguiculata*) flours. International Journal of Food Science and Technology. 2003; (38):403-410
3. Nwabueze TU, Atuonwu AC. Effect of malting African breadfruit (respectively *Treculia Africana*) seeds on flour properties and biscuit sensory and quality characteristics as composite. Journal of Food Technology. 2007; 5(1):42-48
4. Okpala LC, Chinyelu VA. Physicochemical, nutritional and organoleptic evaluation of cookies from pigeon pea (*Cajanus cajan*) and cocoyam (*Xanthosoma sp.*) flour blends. African journal of food, agriculture, nutrition and development. 2011; 11 (6): 5431-5443.
5. Onweluzo JC, Nwabugwu CC. Fermentation of millet (*Pennisetum americanum*) and pigeon pea (*Cajanus cajan*) seeds for flour production: Effects on composition and selected functional properties. Pakistan Journal of Nutrition 2009; 8 (6): 737-744
6. Hallén E, İbanoğlu S, Ainsworth P. Effect of fermented/germinated cowpea flour addition on the rheological and baking properties of wheat flour. Journal of Food Engineering 2004;(63):177-184.
7. ICRISAT (International Crops Research Institute for the Semi- Arid Tropics). Annual Report Patancheru, A.P. India; 2008.
8. Tiwari BK., Tiwari U, Mohan RJ, Alagusundaram K. Effects of various pre-treatments on functional, physicochemical and cooking properties of pigeon pea (*Cajanus cajan* L.) *Food Science and Technology*. 2008;14(6):487-495
9. Awadelkareem AM, Muralikrishna G, El- Tinay AH, Mustafa AI. Characterization of tannin and study of *in vitro* protein digestibility and mineral profile of Sudanese and Indian sorghum cultivars. Pakistan Journal of Nutrition. 2009;8(4):469-476.
10. FAO. Traditional Food Plants. Food and Agriculture Organization of the United Nations. Rome; 1988.
11. Ihekoronye AI, Ngoddy PO. Integrated Food Science and Technology for the Tropics, London, Macmillan; 1985.

12. Huang D. Selecting an optimum starch for snack development (online). 2005. Accessed on 18 June, 2008. Available: http://www.foodinnovation.com/pdf/Selecting%20Optimal%20Starch.pdf.

13. Udensi EA, Gibson-Umeh G, Agu PN. Physico- Chemical Properties of some Nigerian varieties of cocoyam. Journal of Science, Agriculture, Food Technology and Environment. 2008; (8):11-14

14. AOAC. Official Methods of Analysis of Association (15th edition) of Official Analytical Chemists, Washington DC; 1990

15. Reddy MB, Love M. The impacts of food processing on the nutritional quality of vitamins and minerals. Advances in Experimental Medical Biology. 1999; (459):99-106

16. Birk Y, Bondi, A, Gestetner B, Ishaya, IA. Thermostable haemolytic factor in soyabeans. *Natural*. 1963;(197):1089-1090.

17. Hudson, BJ, El-Difrawi EA. The sapogenins of the seeds of four lupin species. Journal of Plant Foods. 1979;(3):181-186.

18. Balagopalan C, Padmaja G, Nanda S, Moorthy SN. Cassava in food, feed and industry. CRC Press, Inc, Boca Raton, Florida; 1988

19. Ukpabi UI, Ejidoh JI. Effect of deep oil frying on the oxalate content and degree of itching of cocoyams (*Xanthosoma* and *Colocasia Spp.*) Technical paper presentation at the 5th Annual Conference of the Agricultural Society of Nigeria. Federal University of Technology, Owerri, Nigeria; 1989.

20. Arntifield SS, Ismond MA, Murray ED. Fate of antinutritional factors during miscellization technique. Canadian Institute of Food Science and Technology Journal. 1985; (18):137-140.

21. Al- Numair KS, Ahmed SE. Biological evaluation of karkade (*Hibiscus sabdariffa*) seed protein isolate. Journal of Food, Agriculture and Environment. 2008;(6):155-161.

22. Prabhavat S, Cuptapun Y, Mesomya W, Jittanoonta P, Bunyawisuthi, S. Nutritional evaluation of snacks from corn flour produced by village texturizer. 1991. Accessed on 2 May 2010. Available: http:// www.pikul.lib.ku.ac.th.

23. Ugwu F, Ugwu S. Nutritive quality of sprouted corn with germinated, fermented and dried jackbean (*Canavalis ensiformis*), *Mucuna vulgaris,* pigeon pea (*Cajanus cajan*) and vegetable cowpea (*Sesquipedalis*). Tropentag. 2009;287-290

24. Olaoye OA, Onilude AA, Oladoye CO. Breadfruit flour in biscuit making: effects on product quality. African Journal of Food Science. 2007;(10):20-23

25. Okpala LC, Okoli EC. Nutritional evaluation of cookies produced from pigeon pea, cocoyam and sorghum flour blends. African journal of biotechnology. 2011;10(3):433-438

26. Mercer LP, Watson DF, Ramlet JS. Control of Food Intake in the rat by dietary protein concentration. Journal of Nutrition. 1981;(111):1117-1123

27. De Angelis RC, Takahashi N, Amaral L, Terra IC. Imbalanced protein and appetite. Arq. Gastroent. S. Paulo. 1978;(15):194-198.

28. Nnam, NM. Comparison of the protein nutritional value of food blends based on sorghum, bambara groundnut and sweet potatoes. International Journal Food Science and Nutrition. 2001;(52):25-29

29. Onweluzo JC, Nwabugwu CC. Development and evaluation of weaning foods from pigeon pea and millet. Pakistan Journal of Nutrition. 2009;8 (6):725-730

30. Anyika JU, Obizoba IC, Nwamarah JU. Effect of processing on the protein quality of African yam bean and Bambara groundnut supplemented with sorghum or crayfish in rats. Pakistan Journal of Nutrition. 2009;8(10):1623-1628.

31. PAG. Protein Advisory Group guideline, No 8. Protein-rich mixtures for use as weaning foods. New York FAO/WHO/UNICEF; 1971.

32.	FNB. Recommended daily allowances. National Research Council Publication No. 1694. Washington D.C. National Academy of Science; 1974

33.	Liener IE. Toxic constituents of plant foodstuffs. Academic Press, New York, London; 1980.

Permissions

All chapters in this book were first published in EJFRR, by Science Domain International; hereby published with permission under the Creative Commons Attribution License or equivalent. Every chapter published in this book has been scrutinized by our experts. Their significance has been extensively debated. The topics covered herein carry significant findings which will fuel the growth of the discipline. They may even be implemented as practical applications or may be referred to as a beginning point for another development.

The contributors of this book come from diverse backgrounds, making this book a truly international effort. This book will bring forth new frontiers with its revolutionizing research information and detailed analysis of the nascent developments around the world.

We would like to thank all the contributing authors for lending their expertise to make the book truly unique. They have played a crucial role in the development of this book. Without their invaluable contributions this book wouldn't have been possible. They have made vital efforts to compile up to date information on the varied aspects of this subject to make this book a valuable addition to the collection of many professionals and students.

This book was conceptualized with the vision of imparting up-to-date information and advanced data in this field. To ensure the same, a matchless editorial board was set up. Every individual on the board went through rigorous rounds of assessment to prove their worth. After which they invested a large part of their time researching and compiling the most relevant data for our readers.

The editorial board has been involved in producing this book since its inception. They have spent rigorous hours researching and exploring the diverse topics which have resulted in the successful publishing of this book. They have passed on their knowledge of decades through this book. To expedite this challenging task, the publisher supported the team at every step. A small team of assistant editors was also appointed to further simplify the editing procedure and attain best results for the readers.

Apart from the editorial board, the designing team has also invested a significant amount of their time in understanding the subject and creating the most relevant covers. They scrutinized every image to scout for the most suitable representation of the subject and create an appropriate cover for the book.

The publishing team has been an ardent support to the editorial, designing and production team. Their endless efforts to recruit the best for this project, has resulted in the accomplishment of this book. They are a veteran in the field of academics and their pool of knowledge is as vast as their experience in printing. Their expertise and guidance has proved useful at every step. Their uncompromising quality standards have made this book an exceptional effort. Their encouragement from time to time has been an inspiration for everyone.

The publisher and the editorial board hope that this book will prove to be a valuable piece of knowledge for researchers, students, practitioners and scholars across the globe.

List of Contributors

K. B. Filli
Department of Food Science and Technology Federal University of Technology Yola, PMB 2076 Yola, Adamawa State, Nigeria and SIK – The Swedish Institute for Food and Biotechnology Gothenburg, Sweden

I. Nkama
Department of Food Science and Technology University of Maiduguri PMB 1069 Maiduguri, Borno State, Nigeria

V. A. Jideani
Department of Food Technology, Cape Peninsula University of Technology, Bellville 7535, Cape Town, South Africa

U. M. Abubakar
Department of Food Science and Technology University of Maiduguri PMB 1069 Maiduguri, Borno State, Nigeria

Alfonso Clemente
Department of Physiology and Biochemistry of Nutrition, Estación Experimental del Zaidín (CSIC), Profesor Albareda 1, 18008 Granada, Spain

Firew Lemma Berjia
Division of Epidemiology and Microbial Genomics, The National Food Institute, Technical University of Denmark, Mørkhoj Bygade 19, 2860 Søborg, Denmark

Rikke Andersen
Division of Nutrition, The National Food Institute, Technical University of Denmark, Mørkhoj Bygade 19, 2860 Søborg, Denmark

Jeljer Hoekstra
National Institute of for Public Health and the Environment (RIVM) Bilthoven, The Netherlands

Morten Poulsen
Division of Toxicology and Risk Assessment, The National Food Institute, Technical University of Denmark, Mørkhoj Bygade 19, 2860 Søborg, Denmark

Maarten Nauta
Division of Epidemiology and Microbial Genomics, The National Food Institute, Technical University of Denmark, Mørkhoj Bygade 19, 2860 Søborg, Denmark

Janina Willers
Institute of Food Science, Leibniz University Hannover, Hannover, Germany

Stefanie C. Plotz
Institute of Food Science, Leibniz University Hannover, Hannover, Germany

Andreas Hahn
Institute of Food Science, Leibniz University Hannover, Hannover, Germany

Laura Massini
School of Food Science and Environmental Health, Dublin Institute of Technology, Cathal Brugha Street, D1, Dublin, Ireland

Daniel Rico
Agro Technological Institute of Castilla y Leon (ITACYL), Government of Castilla y Leon, Finca Zamadueñas, Valladolid, Spain

Ana Belen Martín Diana
Agro Technological Institute of Castilla y Leon (ITACYL), Government of Castilla y Leon, Finca Zamadueñas, Valladolid, Spain

Catherine Barry-Ryan
School of Food Science and Environmental Health, Dublin Institute of Technology, Cathal Brugha Street, D1, Dublin, Ireland

Bastiaan J. Venhuis
RIVM - National Institute for Public Health and the Environment, P.O. Box 1, NL-3720 BA Bilthoven, The Netherlands

Dries de Kaste
RIVM - National Institute for Public Health and the Environment, P.O. Box 1, NL-3720 BA Bilthoven, The Netherlands

Mahdieh Iranmanesh
Department of Food Science and Technology, Science and Research Branch, Islamic Azad University, P.O. Box 14515.775, Tehran-Iran

Hamid Ezzatpanah
Department of Food Science and Technology, Science and Research Branch, Islamic Azad University, P.O. Box 14515.775, Tehran-Iran

Naheed Mojgani
Biotechnology Department, Razi Vaccine and Serum Research Institute, Karaj-Iran

Mohammad Amir Karimi Torshizi
Department of Poultry Science, Faculty of Agriculture, Tarbiat Modares University, Tehran-Iran

Mehdi Aminafshar
Department of Animal Science, Science and Research Branch, Islamic Azad University, P.O. Box14515.775, Tehran-Iran

Mohamad Maohamadi
Department of Food Science and Technology, Science and Research Branch, Islamic Azad University, P.O. Box 14515.775, Tehran-Iran

F.C.K. Ocloo
Biotechnology and Nuclear Agriculture Research Institute, Ghana Atomic Energy Commission, P.O. Box LG 80, Legon, Ghana

A. Adu-Gyamfi
Biotechnology and Nuclear Agriculture Research Institute, Ghana Atomic Energy Commission, P.O. Box LG 80, Legon, Ghana

E. A. Quarcoo
Biotechnology and Nuclear Agriculture Research Institute, Ghana Atomic Energy Commission, P.O. Box LG 80, Legon, Ghana

Y. Serfor-Armah
National Nuclear Research Institute, Ghana Atomic Energy Commission, P.O. Box LG 80, Legon, Ghana

D. K. Asare
Biotechnology and Nuclear Agriculture Research Institute, Ghana Atomic Energy Commission, P.O. Box LG 80, Legon, Ghana

C. Owulah
Biotechnology and Nuclear Agriculture Research Institute, Ghana Atomic Energy Commission, P.O. Box LG 80, Legon, Ghana

Siti Nurdjanah
Department of Agriculture Product Technology, University of Lampung, Lampung 35145,Indonesia, and a former Ph.D Research Student at Food Science and Technology, School of Chemical Engineering, Faculty of Engineering The University of New South Wales, Sydney, NSW 2052 Australia

James Hook
NMR Facility, The University of New South Wales, Sydney, NSW 2052 Australia

Jane Paton
Faculty of Science, The University of New South Wales, Sydney, NSW 2052, Australia

Janet Paterson
Food Science and Technology, School of Chemical Engineering, Faculty of Engineering-The University of New South Wales, Sydney, NSW 2052, Australia

Emmanuel K. Gasu
School of Nuclear and Allied Sciences Atomic Campus University of Ghana, Ghana

Victoria Appiah
School of Nuclear and Allied Sciences Atomic Campus University of Ghana, Ghana

Abraham Adu Gyamfi
Biotechnology and Nuclear Agriculture Research Institute, Ghana Atomic Energy Commission, Ghana

Josehpine Nketsia-Tabiri
Biotechnology and Nuclear Agriculture Research Institute, Ghana Atomic Energy Commission, Ghana

M. O. Oluwamukomi
Department of Food Science and Technology, Federal University of Technology, Akure, Nigeria

I. A. Adeyemi
Bells University of Technology, Ota, Nigeria

L. C. Okpala
Department of Food Science and Technology, Ebonyi State University, Abakaliki, Nigeria

O. O. Ekwe
Department of Animal Science, Ebonyi State University, Abakaliki, Nigeri